Knowledge, Inequality and Growth in the New Economy

Knowledge, Inequality and Growth in the New Economy

Richard Nahuis

CPB Netherlands Bureau for Economic Policy Analysis and Utrecht School of Economics, Utrecht University, The Netherlands

Edward Elgar

Cheltenham, UK • Northampton, MA, USA

Published by
Edward Elgar Publishing Limited
Glensanda House
Montpellier Parade
Cheltenham
Glos GL50 1UA
UK

Edward Elgar Publishing, Inc.
136 West Street
Suite 202
Northampton
Massachusetts 01060
USA

A catalogue record for this book
is available from the British Library

Library of Congress Cataloguing in Publication Data

Nahuis, Richard.
Knowledge, inequality and growth in the new economy / Richard Nahuis.
p. cm.
Includes bibliographical references and index.
1. Income distribution. 2. Wages—Effect of education on. 3. Wages—Effect of technological innovations on. 4. Technological innovations—Economic aspects. 5. Economic development. I. Title.

HC79. I5N34 2003
331.2'2—dc21

2003048537

ISBN 1 84376 323 0

Printed and bound in Great Britain by MPG Books Ltd, Bodmin, Cornwall

Contents

PART II THEORY

Acknowledgements

The book in front of you is the product of four and a half years of research carried out at the Department of General Economics and CentER of Tilburg University and CPB Netherlands Bureau for Economic Research.

Without the help and encouragement of others, the book would have looked very different or even have been non-existent. The latter would certainly have been the case without my supervisor Theo van de Klundert. His lectures made economics interesting and challenging. Moreover, he co-authored Chapter 8 and his encouragement and skilful advice shaped this book to a considerable extent. I think that he found a midway between giving me freedom to pursue research ideas that interested me and ensuring that I completed the book. Thanks for all that.

Knowledge spillovers are necessary for growth. Some colleagues (and friends) were important sources for knowledge spillovers during the research and in this way they considerably contributed to this book. I would like to express my appreciation for their influence.

Henri de Groot, always emphasizing empirical relevance, gave valuable comments on all chapters, co-authored chapters 5 and 7 and was, above all, open-minded in the numerous discussions on diverse subjects. Chapter 7 is also co-authored by Erik Canton and he is acknowledged for initiating that project.

Sjak Smulders, encouraging me to provide a sound intuition behind any result or equation, was always helpful when it came to technical difficulties. I learned a lot from him in a joint project which resulted in Chapter 6.

Bas Jacobs' enthusiasm is catching and an asset for any research organisation. It was pleasant working with him on Chapter 4.

Arjan Lejour was a pleasure to work with on what has become Chapter 9.

Paul Tang, always questioning whether a project was interesting, was a 'badger' who was an inspiratory colleague at CPB.

Jan Boone, urging me to make things simple, is acknowledged for the many discussions on different topics that appear in this book.

In addition to those already mentioned, many other people made valuable comments on the various chapters.

For a third of the time that I was working on this book I was based at the International Economic Analysis unit of CPB. I am grateful to the CPB for its hospitality and the pleasant atmosphere. I want to acknowledge Hans Timmer, head of the unit at that time, for the trust he expressed in giving me freedom to carry out my research, and for the helpful conversations we had. The colleagues in the unit, Nico, Arjen, Ben, Ton, Johannes, Gerard, Guido and Kathy made it a friendly environment.

I met a number of Ph.D. students in Tilburg, all of whom made my time very enjoyable. Jeroen van de Ven shared my office in the final year and worked with me on smaller projects. Marco, Edward, Bas (3), Gijs, Mila, Willem, Jan, Theo, Michèle and many others made coffee breaks, lunches and 'AIO-uitjes' pleasant.

I enjoyed writing this book and the people mentioned above all contributed to that. However, my friends and family created the right environment. It is hard to explain how important the Friday night 'discussions' with Ward and Patrick were. Many other friends were and are also important. I hope they all know, but still I want to express it explicitly to Manon, Marcel, Frank and Susan. There is no need to explain that Inge became important..... I want to thank my mother, who always encouraged me and worked hard to facilitate all I did. Thanks.

Finally, I acknowledge the following publishers for granting me permission to use the material in several chapters from a variety of published papers.

Chapter 4: Reprinted from ECONOMICS LETTERS, Vol. 74, Bas Jacobs and Richard Nahuis, "A General Purpose Technology Explains the Solow

Paradox and Wage Inequality", pages 243–50, 2002, with permission from Elsevier. Chapter 5: Reprinted from EUROPEAN JOURNAL OF POLITICAL ECONOMY, Vol. 18, Erik Canton, Henri de Groot and Richard Nahuis, "Vested Interests and Resistance to Technology Adoption", pages 631–52, 2002, with permission from Elsevier.

Chapter 6: Reprinted from JOURNAL OF ECONOMIC GROWTH, Vol. 7, Richard Nahuis and Sjak Smulders, "The Skill Premium, Technological Change and Appropriability", pages 137–56, 2002, with permission from Kluwer.

Chapter 7: Reprinted from JOURNAL OF ECONOMICS, Vol. 76, Henri de Groot and Richard Nahuis, "Optimal Product Variety and Economic Growth: The Trade-off between Internal and External Economies of Scale", pages 1–32, 2002, with permission from Springer-Verlag.

Chapter 8: Reprinted from ECONOMIC MODELLING, Vol. 15, Theo van de Klundert and Richard Nahuis, "Economic Development and Trade in the World Economy", pages 287–305, 1998, with permission from Elsevier.

Richard Nahuis, Den Haag, July 2003.

PART I

Introduction

1. Knowledge and economic growth: General introduction and outline

1. INTRODUCTION

The increase in material well-being experienced in the last two centuries has been tremendous. For most eras one can point at some major technological breakthroughs that were driving forces of growth. It is likely that the invention of the steam engine, electricity, and the internal combustion engine have been very important for growth. It seems that growth was relatively rapid in some periods, related to major technological breakthroughs, whereas it was slow in others. The dominant technology in our era – information and communication technology – does not seem to drive productivity growth to new heights. Inequality, however, seems to be higher than it has been for decades. This book is concerned with factors determining the size of the economic cake as well as the division of the cake.

The first question this book deals with is how do we grow? Do we grow by doing or using more of the same? Or by doing the same more efficiently? Doing more of the same is inevitably bounded by the limits to the available resources. Increasing production by increasing the number of farmers and acres of production area is inevitably limited by the availability of fertile land. Analogously, using more of the same – that is, giving a factory worker more and more machines to work with – ultimately runs into diminishing returns. It is, however, hard to imagine a natural limit on doing the same more efficiently. Putting more and more ideas and knowledge in the machine that the factory worker works with is not likely to run into diminishing returns that are so prevalent elsewhere in economic life.

What these simple observations actually tell is a little history of economic science: from Ricardo's diminishing returns in agriculture, to Solow's production possibilities with constant returns to flexible combinations of capital and labour, onto the endogenous growth theories with constant returns to a (single) reproducible factor like knowledge. Solow's (1956) contribution – which initiated an earlier round of intense theorizing in the field of economic growth[1] – disentangled output growth in growth of production factors and a residual. That is, separating 'more of the same' from 'the same more efficiently'. The latter, now known as Solow's residual, is identified with an outward shift of the production possibilities curve, which is defined as technological progress. Solow (1957) showed that 'his' residual was a very important factor in explaining long-run growth. In Solow's model, however, technological progress is taken as an exogenous constant. Given the fact that the unexplained part is so important and that the sophistication of the production processes differ substantially across firms and countries, Solow's framework is not particularly comprehensive.

The observation that total factor productivity is not equal everywhere and that there is no clear tendency towards convergence motivated a whole new round of work on growth theory. Solow's residual is endogenized by Romer's (1986) seminal contribution. Endogenous growth theory attempts to explain the factor that Solow (1957) had shown to be important: technological progress.[2]

This book deals with growth driven by technological change. The first concern of this book is why technology differs within and across countries. To that aim, technological change and growth will be analysed in its two appearances. First, technological progress manifests itself as an incremental process. The incremental process of productivity growth will be analysed from the angle that technology and knowledge are to a large extent specific. That is, knowledge and ideas are not available to everybody, but are specific to firms. Taking this perspective on technology, the tension between specificity of knowledge and knowledge spillovers comes to the fore immediately. Second, technological progress manifests itself in major technological breakthroughs, so-called General Purpose Technologies (GPTs). Studying the unevenly arriving GPTs adds additional insight, as the induced major adjustments have consequences for both growth and income distribution.

Though growth is telling the story on increasing income levels, the distribution of the gains is of great importance too. Samuelson (1964, p. 773[3]) argues that: 'With the advance of technology and the piling up of a larger stock of capital goods, it would take a veritable miracle of the devil to keep real wages of men from being bid ever higher with each passing decade.' Despite the fact that Samuelson did not distinguish between high- and low-skilled workers, his statement emphasizes how remarkable the experience of low-educated workers has been; they faced declining real wages during two decades.[4] To what 'devil's miracle' can this observation be ascribed? This is the second concern of this book.

The remainder of this chapter does three things. First, the 'specificity' perspective on growth and knowledge, that shows up recurrently in the book, is defined more precisely and the choice for this perspective is motivated. Moreover, as in most chapters the emphasis lies in the application of models to specific (empirical) questions, we will emphasize here the differences between the models developed and their contribution to the general theory of growth. The second part of the chapter discusses two empirical puzzles that will reappear in several chapters. Third, a chapter-oriented reading guide is provided at the end of this chapter.

2. SPILLOVERS AND TECHNOLOGICAL SPECIFICITY IN ENDOGENOUS GROWTH

2.1 Why does productivity differ?

A major concern motivating the research reported in this book is the question of why technology differs within and across countries. Here we explore which answers endogenous growth theory gives to this question. To be clear, the aim of the next section is not to review the complete literature, nor to say the last word on which explanation for international productivity differences is the most plausible. The plausibility of the answers will be discussed to motivate the perspective on growth and knowledge applied in the remainder of the book. In sorting out what is plausible we evaluate explanations on whether they contradict the following facts. (1) Knowledge spillovers occur, but are incomplete or imperfect (Jaffe, Trajtenberg and Henderson, 1993). (2) Knowledge spillovers are not mechanically nor automatically (see Eaton and Kortum, 1996, Cohen and Levinthal, 1989, and Benhabib and Spiegel, 1994). (3) Factor endowments, like skilled and unskilled labour, do not differ tremendously in relative terms between OECD countries. (4) Implementation costs outweigh invention costs by a factor 20 to 30 (see Jovanovic, 1997). (5) Finally, growth rates that increase on a one-to-one basis with the number of people in the economy (a so-called scale effect) are implausible.

2.2 Some answers

Every answer given to the question on differences in technology applied across countries essentially has to negate the extreme (version of the neo-classical) perspective that technology is freely and equally available everywhere. Hence the answer must be some variation of: ‘technology is not freely available or not costlessly and equally efficiently implementable everywhere’. This answer implies that one single technology concept is insufficient to deal with the question. Hence, before specifying the answers further we need to spend some time on language.[5]

Knowledge concepts

We distinguish two levels: knowledge used in the production of goods and knowledge used in the production of knowledge. The first, which is actually implemented technology, or total factor productivity (further, TFP) is a well-defined concept. It is, however, fruitful to formulate it in the terms used throughout this book: it is the efficiency with which to work with blueprints (or patents) and concepts that have been developed (in the remainder we also refer to this productivity factor as *productive knowledge*).[6] In principle productive knowledge is non-rival, but it can be excludable[7] due to patents or secrecy, or it can simply be valueless for others.

That productive knowledge (TFP) differs across agents can be due to differences in information that a specific agent has available for implementation in production and due to differences in inputs available for R&D. This information set is defined as *useable or available knowledge*. It includes: the agent's tacit knowledge and experience, the laws of physics, and ideas and techniques developed by others, known to the agent, which can be used in the development of new knowledge. Useable knowledge is non-rival and usually assumed to be non-excludable. In Solow's model all knowledge is non-excludable. In, for example, Romer (1990) and Grossman and Helpman (1991) useable knowledge for R&D is non-excludable. Here, we argue that the knowledge base for R&D is also excludable (to some extent). Hence, we will argue that this knowledge stock can be specific too, for example due to secrecy. Therefore, effort is required to obtain useable knowledge.[8]

The sum of all information (all there is in storage devices and in people's minds but not necessarily *available* for a specific agent) is simply referred to as *knowledge*. Figure 1.1 illustrates this classification where productive knowledge is a subset of useable knowledge.

Knowledge specificity or technological specificity follows from the fact that many ideas are not codified, weakly defined, and not directly applicable in other environments (for an elaboration see Dosi, 1988 and Smulders, 1994). Specificity is an important concept in thinking about knowledge, growth and development. Moreover, useable knowledge in non-specific in 'benchmark' growth models (Romer, 1990, and Grossman and Helpman, 1991). For these reasons an explicit definition might be helpful: knowledge is specific if it differs 'substantially' in value or usefulness between otherwise seemingly similar environments.[9]

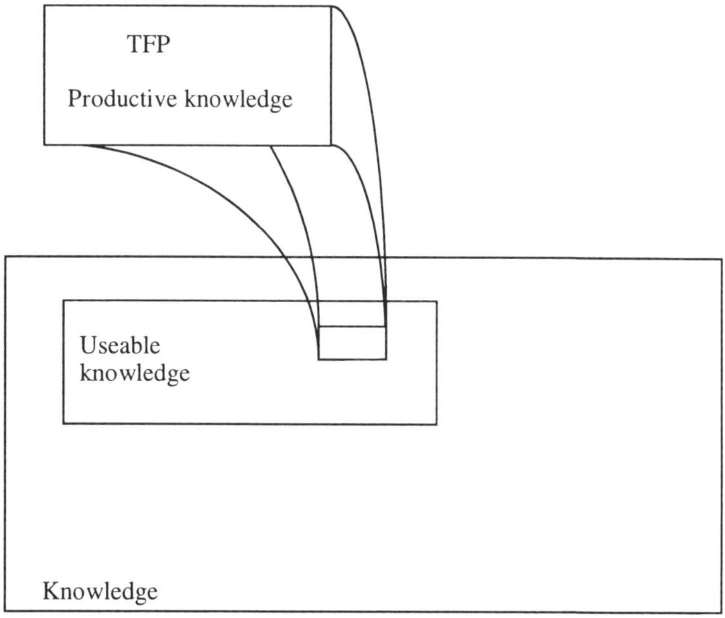

Figure 1.1 Knowledge concepts

On productivity differences

Now we are fully equipped to analyse explanations for productivity differences. The explanation this book explores in detail is easily expressed in the terms introduced above. To explain productivity differences we assume that knowledge that is developed within a certain environment is not useful in other environments, hence knowledge is specific to the environment that developed the knowledge. We focus on firms as agents who create new knowledge. Firm-specific knowledge implies that the development of productive knowledge yields mostly useable knowledge for the firm itself. The approach thus emphasizes the cumulative nature of the research process. This is not to say that spillovers are absent – innovation generally creates more general concepts useful to other firms that are not kept secret perfectly – but to emphasize that knowledge is to a large extent tacit. So, differences in productivity across and within countries are explained by different histories of firms. Hence, the observation that different countries are productivity leaders in different industries (and may lose productivity leadership over time) is explained by the tension between specific knowledge accumulation and knowledge spillovers. This explanation, which is a variation of the localized-technology approach,[10] is referred to in the remainder as the specific-knowledge approach.

A related explanation for which the evidence is favourable is the idea that knowledge does not easily cross borders; that is, the useable knowledge stock for R&D is limited to domestically developed knowledge. This explains why productivity differs across countries (Grossman and Helpman, 1991). Knowledge spillovers do occur but are generally incomplete, and hence explanations based on differences in useable knowledge are plausible. To account for the differences observed by arguing that spillovers are incomplete is one thing, but an explanation for the incompleteness is another. We do this by applying the specific-knowledge perspective.

Different societies have different attitudes to 'the new,' as argued by Mokyr (1990). Technological change is introducing and testing novelties. Therefore, technologies might differ simply because the implementation of best practice productive knowledge is outruled by law. Think, for example, of the opposition to genetic engineering. This perspective does not rely on lack of knowledge flows or differences in useable knowledge, nor is it implausible for the OECD countries (for evidence, see Baily and Gerschbach, 1995). In Chapter 5, we explore why some technologies (like GPTs) are more likely to meet resistance than others.[11]

An explanation that we do not emphasize is that productivity differences across countries are due to differences in the quality of labour (Lucas, 1988). With mobile physical capital, a capital flow from capital-abundant countries to countries with low capital-labour ratios (poor countries) would tend to equalize income. However, Lucas argues that labour quality in LDCs is so low that marginal products of capital are already equalized at the prevailing capital-labour ratios. The argument goes as follows: there are differences in productive knowledge (productivity) that endure, as productive knowledge is tied to immobile people. We do not embrace this perspective, as differences in education between members of the OECD are not very large.

2.3 Knowledge specificity and spillovers: Models of endogenous growth

In this book, we construct models where firms do in-house R&D to improve their product which they sell on a monopolistically competitive market (see Smulders and Van de Klundert, 1995, Peretto, 1996). The models developed focus on different aspects of the specificity and excludability of the useable knowledge stock for R&D. Knowledge spillovers are convincingly found in the literature; references are provided in subsequent chapters. In all models employed in the remainder of this book, spillovers play an important role. Two different perspectives are taken. First, Chapters 6 and 7 assume that stocks of useable knowledge accumulated by firms' research overlap (the shaded part in Figure 1.2). That is, the stock of accumulated knowledge of one firm spills over to another firm and thus increases that firm's stock of useable knowledge. Hence, developed knowledge is only specific to a limited extent. Figure 1.2 provides a schematic representation of two firms' knowledge stocks and research activities.

Second, Chapters 3 and 4 explicitly recognize that knowledge useable for research is specific to some extent. The consequence is that expanding the knowledge base for research (useable knowledge), by assimilating or adopting ideas, is a costly activity too (see Rustichini and Schmitz, 1991, for an early contribution recognizing this second dimension of R&D). Figure 1.3 shows the two activities of a single firm. Note that research generates spillovers to the knowledge stock that can be assimilated by other firms.

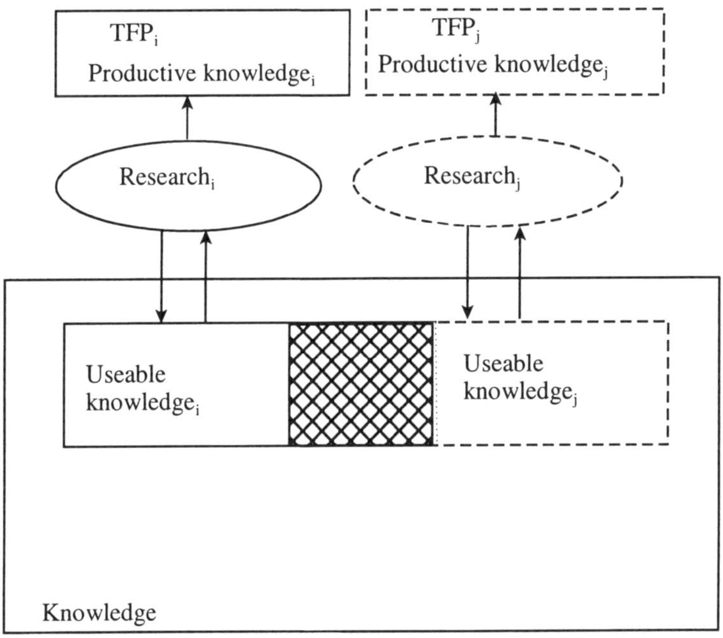

Figure 1.2 Scheme of the models in Chapters 6 and 7

Although this book does not literally apply an adoption model – except for Chapter 5 – an adoption interpretation stresses the duplicationary character of the R&D process. That is, ideas and spillovers are to be implemented and adjusted to specific circumstances in *every* firm, building on its own experience.[12] A set-up where each firm has to invent and/or implement ideas is not vulnerable to the scale-effect critique. Scale is the number of agents (with their endowments) in the economy (Jovanovic, 1997). The scale-effect critique is the fact that the growth rates increase in the seminal growth models as the scale of the economy increases. The prediction is, in some forms, at odds with the data (Jones, 1995a).[13] The growth rates in models constructed in this book increase only if the number of skilled workers per firm increases.[14]

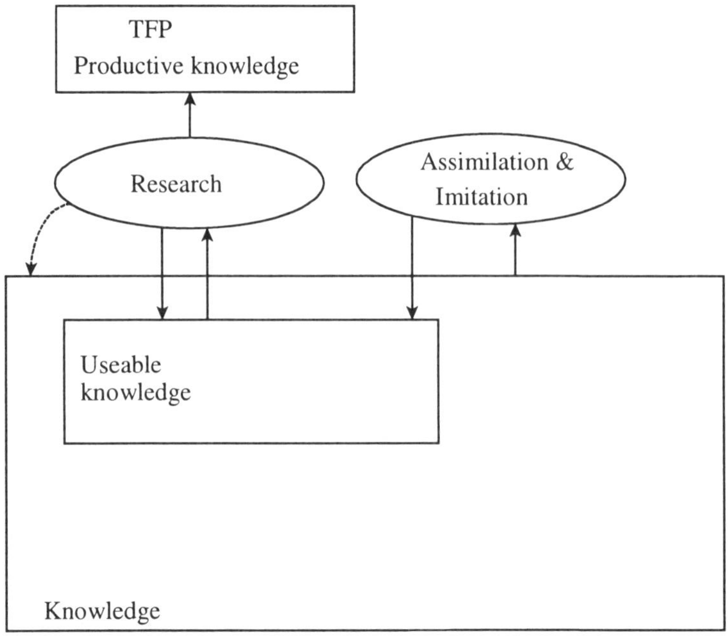

Figure 1.3 *Scheme of the models in Chapters 3 and 4*

The perspective on growth set forth above not only serves the goal of shedding light on the complex process that technological change is; it also provides a different perspective on some empirical puzzles. A discussion of the empirical puzzles, and the contribution of (the perspective of) this book in solving them, is a task to which we turn in the next section.

3. EMPIRICAL PUZZLES: WHAT DOES THE SPECIFICITY PERSPECTIVE ADD?

Two empirical puzzles show up recurrently in this book. First, it is well documented but ill understood that the structure of wages has changed dramatically in recent years. Second, a seemingly revolutionary development in information and communication technology has not paid off in terms of higher growth rates.[15] In this section, we will discuss the facts behind the puzzles, the explanations offered in the literature, and the position taken in the different chapters of this book.

3.1 The wage-inequality puzzle

The structure of wages in the US and UK changed remarkably in the 1970s and 1980s.[16] Three facts stand out. First, the education-wage premium decreased in the 1970s and increased in the 1980s. Second, the return to experience in the labour market increased in the 1980s. Third, residual wage inequality – the variation in wages left after accounting for observable characteristics – increased in the 1970s and 1980s.

Both the first and the second observation say that the premium for 'skills' increased. In the theoretical chapters to follow, the discussion does not distinguish between skill obtained through formal education or through experience.[17] Indicative for the development of the skill premium is the wage difference between non-production (skilled) and production (unskilled) workers. Table 1.1 shows that the non-production/production wage differential in the US and UK declined or was stable from the early 1970s to the early 1980s and increased thereafter. The relative supply of skilled workers increased throughout that period.

The question that arises then is: why was the skill premium stable or declining in the 1970s while it increased in the 1980s despite a continuing increase in the relative supply of skilled workers in *both* periods? The latter part of this question gets most attention in the remainder of the book: why did the skill premium increase in the 1980s? No attempt is made to explain the fact that the skill premium did not dramatically fall throughout the twentieth century despite the continuing increase in the relative supply of skilled workers.[18] Hence, the *steady* increase in demand for skill that has to be going on is of no concern to us.[19] Therefore, read in the remainder 'an *acceleration* in the increase of *relative* demand for skill' when the brief 'increase in demand for skill' is used.

Table 1.1 Labour market data and R&D intensity[a]

	1973	1977	1981	1985	1989
	Non-production/production wage differentials				
US	1.553	1.531	1.532	1.559	1.632
UK	1.316	1.292	1.340	1.366	1.470
Germany					
	Relative supply of college-educated workers[b]				
US	10.8		16.6		21.5
UK	8.0		12.0		18.3
Germany	6.0		7.4		9.4
	R&D intensity (% of value added)				
US	6.3	6.2	7.7	9.7	8.7
UK	4.3	4.6	6.4	6.2	6.0
Germany	3.2	3.7	4.3	5.2	5.5

Notes: [a] Unless indicated differently: for source see Machin and Van Reenen (1998).
[b] Actual years: 1970, 1980, 1989. Percentage of the population. For details see source.
Source: OECD *Employment Outlook*, 1993.

Figure 1.4 illustrates this point (it is a stylized representation of Fig 11.5 in Murphy et al., 1998). The vertical axis indicates the deviation from a constant skill bias in technological change that is necessary to replicate the observed pattern in the skill premium. The implication of the figure is that in the 1980s, given the changes observed in the relative supply of skilled labour, 'something' alongside the steady bias in technological change caused the demand for skill to rise. What change or which shock is responsible for this acceleration in demand?

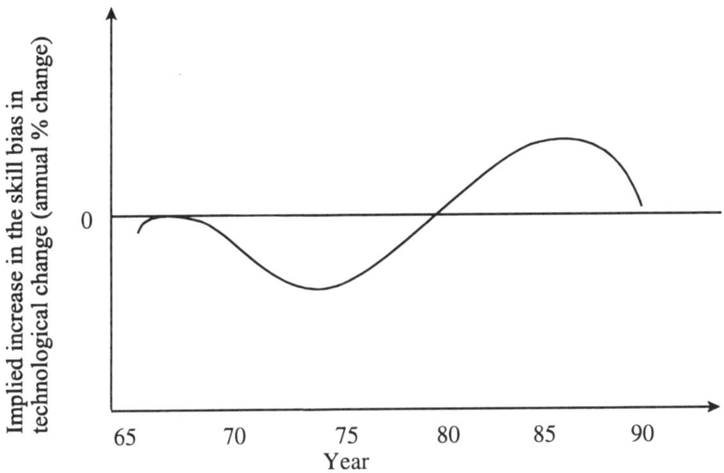

Figure 1.4 *Deviation from a constant skill bias*

Explanations

Four explanations for the increased demand for skill are proposed (Chapter 2 reviews the literature on this issue). First, a more intense trade relation with low-wage countries can cause a relative deterioration in the position of unskilled workers. The hypothesis is that decreased trade barriers and lower transport and communication costs make the unskilled workers' wages in OECD countries more sensitive to the downward pressure exerted by the vast reservoir of unskilled workers in non-OECD countries (see Chapters 8 and 9). Second, changes in the supply of skilled workers are a possible explanation. This explanation comes in different forms. One is that due to incompetent high-schools the relative quality of unskilled workers has deteriorated and hence unskilled workers' relative wage tends to fall. Another form is that, with a constant bias in the demand for skill, a deceleration of the supply of skill might be to blame. Third, changing institutions – think of minimum-wage laws – might explain the observed wage pattern. The final hypothesis is that biased technological change –

many commentators think of the computer revolution – is causing the relative wage decline of unskilled workers (see Chapters 3 and 4).

We zoom in on the biased-technology explanation for a moment. Two things need to be clarified. First, as the relation between technology and inequality depends importantly on the substitutability of skilled and unskilled workers, we discuss two possible assumptions. Second, we discuss what bias in technological change is able to generate inequality.

First, we should answer the question: are skilled workers more efficient equivalents of unskilled workers or are they 'different'? There are two reasons for not choosing the perspective that we can add up efficiency units of labour.[20] First, it is hard to imagine that *any* number of unskilled workers can replace an Einstein in research. Second, in a model that assumes perfect substitution of workers, after correction for education/efficiency, general equilibrium does not affect relative wages. Hence, such a model is incapable of explaining variation in relative wages accompanying long swings in growth.[21]

Second, what is the nature of technological change that can explain inequality?[22] Technological progress might be sector-biased. That is, sectors employing relatively intensively skilled workers experience faster technological progress. This explanation is explored in Chapters 2 and 9. Technological change can also be biased towards skilled workers: the so-called factor bias (for an example, see Chapter 6).[23] Finally, it is well established that skilled workers have a comparative advantage in technology adoption and innovation, see Bartel and Lichtenberg (1987). This comparative advantage of skilled workers leads to inequality if the productivity of research or adoption increases. Hence, technological revolutions might increase skilled workers' productivity and therefore cause changes in inequality (see Chapter 3).[24]

Discussion: one explanation, any explanation, or a non-explanation?

An extensive body of literature, which is reviewed in the next chapter, offers evidence on the relevance of the different explanations. The majority of the literature argues that after scrutiny of the evidence for the first three explanations – trade, supply and institutions – nothing is left but to conclude that biased technological change is causing inequality. This book contributes to this debate in several ways. First, the next chapter defends the proposition that it has not convincingly been shown that trade is irrelevant in explaining wage inequality. Second, given that it has not been shown that trade is irrelevant, we propose to use an applied general equilibrium model to assess the maximum impact of trade on wage inequality. Trade with low-wage countries is responsible for at most half of the increase in wage inequality (see Chapter 8). This is not the same as showing that trade matters, but the opposite is claimed to be wrong. What is not explained by trade should be explained by something else: biased technological change. Having accounted for 'everything else', the residual is caused by the unobservable factor: technological change that is biased. Such an explanation is a non-explanation unless it is made plausible that some change in the economy points indeed at a bias in technological change. A third contribution of this book to the debate on wage inequality is to open the technology black box and to analyse potential mechanisms that translate observed shocks into a bias in technological change. In Chapter 6, a new mechanism from skill supply changes to wage inequality is developed. The alternative explanation for wage inequality is that an acceleration in the supply of skills induces complementary skill-biased investments.[25] Finally, Chapters 3 and 4 explore a mechanism for the skill-bias hypothesis by analysing the relative wage pattern that accompanies the arrival of a GPT like the computer revolution. The computer revolution is an issue to which we turn now.

3.2 The productivity paradox

For every era one can point out a few GPTs, or at least technologies that were or are important. A technology is a GPT if it is widely applicable, requires complementary investments, and has an inherent potential for improvement (for example, due to spillovers that feed back to the GPT).[26] At the turn of the twentieth century electricity was the dominant GPT (see

David, 1991). Even further back in time the steam engine played such a role. Perhaps of a slightly different nature,[27] the development of plastics – or more broadly the chemical revolution – has been a post-WWII GPT. Today, the advances in information and communication technology (IT) are very pervasive (for more examples, see Lipsey et al., 1998).

Figure 1.5 shows the price decline of computers relative to producers' durable equipment (PDE). The fall in the relative price of computers depicted in Figure 1.5 summarizes an annual fall in computer prices of 19 per cent relative to a PDE price increase of 3 per cent annually (Jorgenson and Stiroh, 1995). The upward-sloping curve depicts the accelerating computer investments (scale on rhs). This acceleration and the simple observation that computers, IT or microprocessors (these terms are used interchangeably) were being introduced everywhere gave rise to high expectations for productivity gains. The productivity gains, however, have been disappointing. As Solow (1987) once wrote: 'we see the computer age everywhere except in the productivity statistics'. This is the Solow or productivity paradox.

The productivity paradox has a micro and macro interpretation. The micro interpretation is that investments in computers do not seem to pay off (for evidence, see for example Morrison and Berndt, 1991).[28] The macro interpretation of the paradox is that it is 'disappointing' that the new GPT did not lead to a recovery of productivity growth to levels familiar from the pre-1973 period, see the first line of Table 1.2.[29] However, the hard-to-measure service sector is included in non-farm private business. A closer look at Table 1.2 suggests that the paradox should be interpreted as: productivity growth in the computer-using sectors is disappointing.

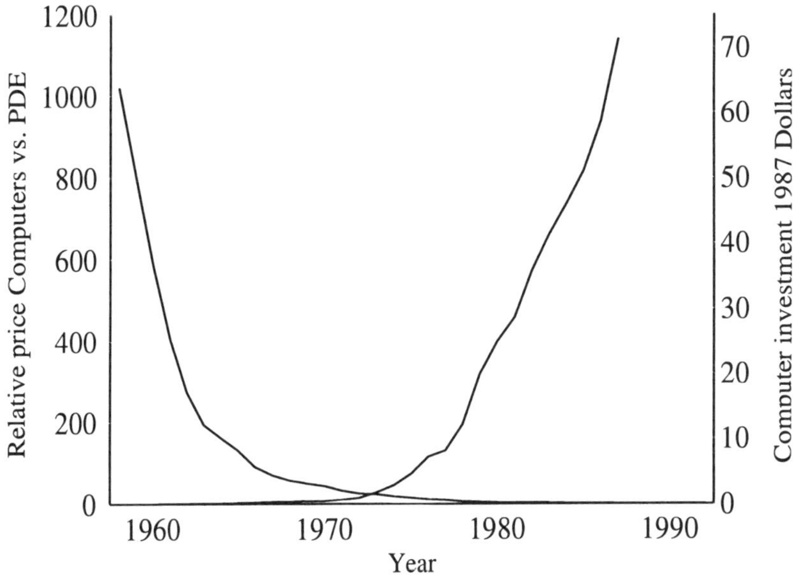

Note: In 1987 the relative price is set at unity.

Figure 1.5 A computer revolution

Explanations[30]

Productivity growth did not show an upsurge despite the massive introduction of computers. To undo this observation from its paradoxical character, a number of explanations have been proposed. One class of explanations is based on adjustment costs. First, it takes time before the truly valuable opportunities of the new GPT are recognized (for anecdotical evidence, see Chapter 3). Second, mis-management with respect to the investment in IT might be relevant (possibly leading to over-consumption of technology). A third (real) explanation is that the size of the computer capital stock is simply not impressive, see Oliner and Sichel (1994).[31]

Table 1.2. The productivity paradox or a time series view of productivity

	1950–72	1972–95	1995–99
	(Annual percentage increase in output per hour)		
Non-farm private business	2.63	1.13	2.15
Manufacturing	2.56	2.58	4.58
Durables	2.32	3.05	6.78
Computer producing	--	17.83	41.70
Computer using	2.23	1.88	1.82
Non-durables	2.96	2.03	2.05

Source: Gordon (2000).

Besides the real explanations, explored further in Chapter 3, there are measurement-error-based explanations (these are explored in Chapter 4). Measurement of productivity requires appropriate measures for output and input. Both are harder to measure now than 50 years ago. Aggregate output data might not fully reflect the value of quality, variety and customization. Moreover, output does not include the produced organizational capital that is related to the adjustment to a new GPT. Inputs are, as a GPT arrives, likely to be overestimated as workers invest in their own human or in firm-specific/organizational capital. Moreover, pecuniary or rent spillovers make it hard to pin down productivity growth.[32]

This book contributes to the effort to undo the productivity figures from their paradoxical character. We do so, first of all, by taking a closer look at the technology black box. Chapter 3 shows that the arrival of a GPT is likely to cause growth to slow down temporarily, once it is acknowledged that public knowledge does not flow freely into the firm but requires assimilation effort.

Second, Chapter 4 shows that productivity growth is likely to be underestimated when a GPT arrives. This chapter also extends the analysis of

Chapter 3 and shows that even the output *level* might go down, along with the growth rate.

A contribution of a slightly different nature is that the next chapters try to convince the reader that increased inequality and disappointing productivity growth might very well *both* be caused by turbulence on the technology front. That is, both inequality and the productivity paradox are driven by a computer revolution.

4. SUMMARY AND READING GUIDE

This section briefly discusses the content of the different chapters in sequential order. Besides discussing what the different chapters do, some remarks are made about the chapters' contributions in the light of this book as a treatise guided by the questions raised in the previous sections. First, however, a reading guide. Readers interested in growth theory are referred to Chapters 3, 5, 6 and 7. Those interested in wage inequality are referred to Chapters 2, 3, 4 and 6. Those interested in the empirics of spillovers should read Chapter 9. Finally, all three elements are integrated in Chapters 8 and 9, which present two applied general equilibrium modelling exercises.

The next chapter reviews the literature on the explanations of the observed increase in wage inequality. The two major explanations, namely trade and biased technological change, are explored by setting out the standard theory and reviewing the evidence for these explanations. Next, deviating explanations are discussed, after which the chapter finishes off with a general discussion. The chapter serves two aims. First, the relations between trade and inequality and between technology and inequality are explained and empirically tested by means of a survey. Second, the chapter shows that the explanation of wage inequality might be based on changes in trade relations but certainly also needs a technology component. It is argued that a technology-based explanation would be more convincing once it has been shown, intuitively appealing, how and why technology is biased towards skilled workers. Chapters 3, 4 and 6 try to fill this gap.

Chapter 3 explores the impact of a GPT in a general-equilibrium framework with skilled and unskilled labour. First, the chapter discusses historical anecdotical evidence regarding the relevance of GPTs, after which some general characteristics are discussed. Next, two types of GPTs are contrasted: an idea-generating GPT and a non-idea-generating GPT. Finally, the relation between the arrival of a GPT and inequality is explored. The chapter makes two points. First, it shows that the recognition of the specificity of knowledge leads to a richer description of the knowledge generation process within firms, as the chapter separates assimilation/imitation from research/innovation. Second, within such a structure it is shown that to explain the productivity paradox and the wage-inequality puzzle one single cause suffices. The arrival of a new GPT leads to lower growth temporarily *and* increased inequality.

Chapter 4 applies a simplified version of the model in Chapter 3. Chapter 4 focuses on the measurement problems that are likely to be relevant in times of a major technology shock, such as a GPT. The chapter makes three points. First, the simplified set-up, as compared to Chapter 3, allows for an analytical corroboration of the transitional dynamics of a GPT shock (derived numerically in Chapter 3). Second, four sources of erroneous measurement in times of major technological change are identified. Third, it is shown that dual growth accounting can provide a solution to overcome the observed measurement problems.

Chapter 5 provides an explanation for observed productivity differences (in a cross-sectional perspective) by exploring how vested interests block the adoption of a GPT. The chapter explains why some societies might turn out to be more productive than others.

Chapter 6 is the final chapter in the theory section that has a positive character. The chapter shows that an alternative explanation for the increase in wage inequality – based on the fact that educational attainment increased – is plausible given that one explicitly recognizes the difference between production and non-production work. Non-production work is indirectly productive and therefore has the character of an investment good. The chapter shows that in a model using the notion of firm-specific knowledge an upward-sloping demand curve for skill might follow. The chapter serves two goals. First, it shows that the firm-specific-knowledge perspective can lead to contrasting results from those obtained applying the perspective where useable knowledge for research is fully public. Second, it shows that an upward-sloping demand curve for skill is only possible if knowledge spillovers are not 'too' substantial, hence emphasizing the tension between specificity and spillovers of knowledge.

Chapter 7 provides a normative analysis of the trade-off between growth and variety in a model that has – to start with – purely rival knowledge. The welfare analysis shows that the market does not deliver the first-best social optimum. Crucial here is the productivity of variety in production versus the competitiveness in the market for intermediate goods. The chapter emphasizes that accumulation of firm-specific knowledge has to some extent a duplicationary aspect. This leads to different results from models where useable knowledge is fully public. The chapter contributes to this book in that it presents what might be called a reference version of the endogenous growth model with firm-specific knowledge. This version of the model discerns only one type of labour. We also explore the implication of allowing for free entry of firms, an aspect largely ignored in the remainder of the book.

Chapter 8 does analyse growth in a Heckscher-Ohlin world, first in theory and then in an applied general equilibrium model. The chapter serves two purposes: first it assesses the role of trade in explaining wage inequality and second a version of the applied general equilibrium without endogenous technological progress is introduced and compared with the standard trade theory.

Chapter 9 builds onto Chapter 8. Endogenous technological progress is introduced in the applied general equilibrium model by introducing knowledge spillovers related to R&D. This chapter analyses the interaction between technology, trade and inequality. The chapter shows that the aggregate perspective used in the analytical chapters can hide interesting aspects related to sectoral reallocation. Trade liberalization might turn out to be lowering growth when knowledge spillovers are tied to trade flows.

NOTES

1. An impression of the excitement growth theory evoked in the 1950s and 1960s is found in the contributions of Stiglitz and Sheshinsky in Diamond (1990).
2. This statement of what endogenous growth theory is about implies almost directly that we focus on technology-based growth models. A less caricatural account of endogenous growth theory is that it tries to overcome the flaws of the neo-classical growth model (which is more than the exogeneity of technological progress). Scott (1989) provides an overview of problems the neo-classical model is unable to explain. A comprehensive overview of all lines followed in 'endogenous growth theory' is found in van de Klundert and Smulders (1992) or Barro and Sala-i-Martin (1995). Endogenous growth theory provided micro-foundations for R&D decisions and put imperfectly competitive market structures in the centre of the analysis (made necessary by the acknowledgement that developing, imitating, or implementing technology is costly).
3. Cited via Murphy et al. (1998).
4. This claim, relevant for the US, will be justified extensively in the next chapter.
5. Discussions about knowledge are numerous (see, for example, Mokyr, 1998); a certain arbitrariness in defining concepts is hardly avoidable.
6. Jovanovic (1997) defines 'the knowledge of how to work the blueprints' (p. 318) as human capital. This terminology is not followed as human capital implies embodiment in persons (a restriction that is too narrow for the approach followed in this book) and is easily confused with skilled workers, see also note 17. A discussion along the lines set out by Jovanovic is provided in Nahuis and van de Ven (1999).
7. Excludability, excluding others from using the productive knowledge, governs appropriation of the returns to R&D investment, which is essential for the incentives facing profit-maximizing firms.
8. For evidence, see Adams (2000).
9. This definition is slightly but importantly different from the definition provided by Caballero and Hammour (1998, p. 724): 'An asset is specific to a relationship to the extent that its value is greater within the relationship than outside.' The important difference is that Caballero and Hammour stress the specificity of an asset in a *relationship*. Quasi-rents generated by a relation-specific asset are, once the investment is made, subject to 'holdup' problems leading to suboptimal investment. We stress that knowledge is specific to a certain firm, but ignore the 'holdup problem' that might potentially arise. In the firm-specific knowledge models discussed in this book, the ownership structure is implicitly defined as shareholders owning the specific knowledge. Probably, however, managers and researchers make a difference in the value of firm-specific knowledge; a difference they could exploit by means of extracting rents from the shareholders. Ignoring this aspect is

similar to the assumption that contracts are complete and enforceable.

10. This approach is based on the idea that the technology is available but useless with the given production technique. That is, a lack of complementary assets leads to low productivity. This perspective explains why a seemingly better technology may actually not be superior in specific circumstances. So the argument is: *productive and useable knowledge* might be available but is valueless given the domestic or local endowments. Localized learning (Atkinson and Stiglitz, 1969, Stiglitz, 1987 and Antonelli, 1995), or appropriate technology, as Basu and Weil (1996) call it, is formalized as learning by doing or innovation that does not shift the *total* isoquant inward, but only the part that is actually used (hence where the initial technology choice was optimal given the factor prices). The important difference between the localized-technology perspective and the specific-knowledge perspective is that the former depends on differences in endowment whereas the latter depends on differences in history and lack of spillovers.
11. See also Holmes and Schmitz (1994).
12. Jovanovic (1997) argues that research spending for adoption outweighs spending for innovation by a factor of 20 to 30. Chapter 7 analyses welfare in a model that has – to start with – purely rival knowledge and hence focuses on the costly duplicationary aspects of innovation.
13. Adoption costs (Jovanovic, 1997) and duplication of research (Young, 1998) are not exactly one and the same thing (Jovanovic is elaborate on this). Alternatively, the scale critique can be circumvented by assuming that the cost of research increases relative to production (see Jones, 1995b and Chapter 6). Evidence, however, that the scale effect prediction *is* consistent with the very long-term data (from 1 million BC to now) is provided by Kremer (1993).
14. This sets the stage for a discussion on free entry and exit of firms. This discussion is postponed till Chapter 7 as in all other chapters the number of firms is fixed; an assumption that fits naturally with the specificity of knowledge (see, de Groot, 1998 for an extensive discussion). For an elaboration on entry, see Peretto and Smulders (2002).
15. With respect to both issues, time is changing the perspective. At the finalizing stage of this book there seems to some extent consensus that wage inequality is largely a 'technological phenomenon' (Chapter 5 explains the technological phenomenon). With respect to the second issue: the US economy did surprisingly well in the late 1990s, giving rise to a debate on the so-called 'new economy'. It remains, however, true that the productivity growth in the computer-using industries remains relatively disappointing; see Table 1.2.
16. The increase in wage inequality in the 1980s got a lot of attention. For references, see Chapter 2.
17. In our perspective, skilled workers have training that causes them to have a comparative advantage in R&D and technology implementation, see Romer (1990). Skilled workers should not be confused with human capital (at least as it is used, for example, in Jones, 1996).
18. Chapter 6 discusses both the 1970s and the 1980s.

19. The relation between factor supplies and factor rewards is not as simple as that suggested here. From the point of view of trade economists, there is no such relation at all. See Chapter 2 for an elaboration.
20. In some chapters it is assumed, for expositional ease, that unskilled workers have zero productivity in research activity and skilled workers have zero productivity in production (Chapters 4 and 6 discuss the sensitivity of the results for this assumption).
21. Bartel and Lichtenberg (1987) show that educated workers have a comparative advantage in implementing new technology. Goldin and Katz (1996) show the relevance of this observation throughout the industrial history.
22. The relation between growth and inequality might run in the opposite direction of causality: from inequality to growth. Persson and Tabellini (1994), for example, argue that inequality is bad for growth, as, in a democracy, a desire for redistribution will lead to (increasingly) distortive taxation. Redistribution might, however, be growth enhancing if it, for example, corrects imperfections that exist in educational investments (for a more extensive critique on the political-economy argument, see Saint Paul and Verdier, 1996). Although in the remainder the focus is entirely on the causal mechanism from growth to inequality, the results should be treated cautiously as (ambiguous) feedback effects from inequality to growth are ignored.
23. Hence, at given relative wages, a more skill-intensive production technique will be chosen. This translates into an output expansion of unskilled-intensive goods. In a closed or large-open economy, this can result in depressed relative wages of unskilled workers; hence increased inequality (see Chapter 2).
24. This mechanism, which will be explored extensively in the remainder, is not entirely distinct from a factor (or sector) bias, as a GPT is technology shock that is beneficial for a certain production factor (or sector).
25. See Chapter 6 and Acemoglu (1998).
26. Chapter 3 defines more precisely what a GPT is, see also Lipsey et al. (1998).
27. Chapter 3 examines two types of GPTs.
28. For contrasting evidence, see Brynjolfsson and Hitt (1997). By using firm-level data for the period 1987–94, they find that IT investments generate 'normal' short-term returns and long-run benefits that exceed those by a factor of 2 to 8.
29. For the productivity slowdown in the early 1970s, many plausible explanations are available, see Romer (1987) and van de Klundert (1999).
30. Based on Chapters 3 and 4 and Brynjolfsson and Hitt (1997, 1998).
31. Though the input share of computers is low, the output elasticity might be higher as computers affect the production process itself and thus TFP (see Brynjolfsson and Hitt, 1997).
32. See Chapter 9 for a detailed discussion on rent spillovers. A variation of the rent-spillovers explanation is given by Brynjolfsson (1996). He argues that a suitable way to measure the 'return' to IT investment is to estimate the generated consumers' surplus.

2. On technology, trade and wage inequality: A survey

1. INTRODUCTION

Declining real wages by 11 per cent in the US in the 1980s![1] Are we living in an age of diminishing prosperity? The decline in real wages is not just experienced by a small fraction of all workers. Less-skilled workers have actually been facing declining wages for a long period, while high-skilled workers experienced increasing real wages.

The populist press connected the rise in inequality in the 1980s immediately with trade with low-wage countries. Obvious explanations are not always correct, hence some academic economists gave a counter-reaction by arguing that the role for foreign trade in this matter is negligible. By doing so, they initiated a lively debate on the cause of the observed decrease of the relative wage of low-skilled workers.

This chapter reviews the literature dealing with the question: 'What caused the decrease in the relative wage of low-skilled workers in the US in the 1980s?' The question implies that the chapter focuses on empirical results and on the US.

The remainder of the chapter is organized along the following lines. The next section more elaborately discusses the facts that instigated the debate. Section 3 analyses the question in a simple supply and demand framework. It is obvious that, as the relative wage *and* the relative supply of high skilled workers increased in the 1980s, the basic question deals with the demand side. Section 4 deals with the analysis of changes in the relative demand for skilled labour in relation to trade. This section's framework is the classroom version of the Heckscher-Ohlin model of trade. Confusing and disappointing results with this analytical framework led many authors to point at biased technological change to explain the changes in relative demand for skilled workers. Section 5 deals with theory and econometric results related to the biased technological change explanation. The modest performance of the classroom trade model directed others to study a deviating or more extended trade model. Section 6 reviews this work. Section 7 presents some deviating approaches, mostly approaches that not have been scrutinized empirically. Section 8 weighs out all approaches and concludes.

2. DOCUMENTING THE FACTS

We first document the facts and postpone the discussion of potential causes. A few prominent trends characterized the US labour market in the 1970s and 1980s. The first is a stagnant development of average real wages. Until 1973 real wages rose rapidly, in the period from 1963 to 1973 there was an increase of 25 per cent! In the period from 1973 to 1989, real wages were roughly constant (Murphy and Welch, 1992). The second development is that *within* narrowly defined groups on the labour market inequality increased.[2] The third is that in clear contrast with the 1970s, the earnings differential between groups with high and low levels of education increased (see Table 2.1). The fourth trend is that educational attainment rose over the 1980s (see Table 2.2). Combining the second and third trend we formulate a fifth trend: overall inequality increased considerably. For a more extensive overview of labour-market trends in the 1980s see Kosters (1994).

This chapter primarily focuses on the increased wage inequality between groups with different educational attainment in the 1980s. Empirical analysis of the earnings inequality produced a bulk of literature. We restrict ourselves to the more recent contributions. For a comprehensive survey of the pre-1990s work, the reader is referred to Levy and Murnane (1992).

Table 2.1 Change in education wage premium in the 1970s and 1980s

	1970s	1980s
United States	-0.08	0.09
Australia	-0.17	0.03
Canada	-0.13	0.03
Germany	NA	0.10
Japan	-0.04	0.00
Netherlands	NA	-0.35
Sweden	-0.09	0.03
United Kingdom	-0.06	0.08

Notes: Five-year average change in the ratio of earnings of college level workers to high-school level workers.

Source: OECD *Employment Outlook* 1993.

Similar changes in education premiums and educational attainment as those described above for the US are found in other countries. Table 2.1 shows education wage premiums for several countries; depicted are the five-year average changes in the earnings ratio of college versus high-school educated. A quick glance at the table shows similar changes in most countries, with the Netherlands and Japan as notable exceptions. Table 2.2 reveals that changes in relative supply in other countries followed a similar pattern to that in the US.

The remainder of this chapter tries to figure out what the explanations are for the facts presented above. Related facts and other trends that might be a clue are the following. (1) The return to experience rose substantially in the 1980s. Thus, this alternative indicator of human capital was also rewarded increasingly (Kosters, 1994). (2) The manufacturing sector shrank substantially. The share of manufacturing employment fell from 26 per cent to 18 per cent between 1971 and 1990. This trend is a common phenomenon in all industrial countries. (3) The gender wage gap declined. (4) Productivity growth slowed down in most of the OECD countries in the early 1970s. The consequence of the slowdown is of course a more modest development of average real earnings. (5) Growth in the trade volume was enormous in the 1970s and slowed down in the 1980s.

Table 2.2 Relative supply of college-educated workers[a,b]

	1970	1980	1989
United States	10.8	16.6	21.5
Germany	6.0	7.4	9.4
Japan	12.0	17.9	22.5
France	5.3	8.3	11.8
United Kingdom	8.0	12.0	18.3

Notes: [a] Percentages of the population.
[b] For details, see source.
Source: OECD *Employment Outlook* 1993

3. SUPPLY, DEMAND AND INSTITUTIONS

A simple framework will be applied for analysing the complex question on the causes of the observed relative wage change.[3] We analyse demand and supply for labour and the institutional settings on the labour market. Here a broad definition of demand shifts is taken. The essential distinction between shifts *of* and *along* the demand curve will be made later on.

Murphy and Welch (1992)[4] focus on relative supply changes and test whether this, in combination with a constant demand structure, fits the facts. What they basically apply is a simple supply and demand framework wherein workers are differentiated by age and education. First they examine the assumption of a stable labour demand structure. For all groups, supply changes should be reflected in opposite wage changes. For the 1963–79 interval this prediction is confirmed. The decreasing college premium in the 1970s is matched by a relatively strong increase in average educational attainment. For the period after that (1980-89) the results indicate that a strong demand shift has been accompanying the slowing supply increases of educated workers. The next step is to assess the nature of the demand shift for this latter period. A stable trend rate in demand change is consistent with the facts. An interpretation where the demand shift is related to unemployment shifts,[5] together with trade penetration, is neither rejected.

To sum up: demographics (or stated differently, supply changes of educated relative to less-educated workers) are important. But to clarify the complete picture, demand changes should be examined. In the remainder no attempt is made to explain the fact that the skill premium did not dramatically fall throughout the twentieth century (and in the 1970s) despite the continuing increase in the relative supply of skilled workers. Hence, the *steady* increase in demand for skill that has to be going on is of no interest to us. Therefore, read in the remainder 'an *acceleration* in the increase of *relative* demand for skill' when the brief 'increase in demand for skill' is used.

As supply shifts cannot bear the sole responsibility for the observed trend, a first step in detecting demand shift causes is made by Katz and Murphy (1992).[6] They attempt to connect jobs with industries. In their analysis, Katz and Murphy distinguish *within*-industry and *between*-industry demand shifts. Within industry shifts are shifts changing relative factor intensities at given relative factor prices; these are related to changing non-labour input prices[7] and non-neutral factor productivity change, for example. Between-industry shifts are those that change the allocation of labour between industries at fixed relative factor prices and are connected with (trade induced[8]) demand shifts or differences between industries in factor productivity growth. Again, besides supply changes, demand changes in favour of the college educated workers are necessary to fit the facts (whether a stable demand trend does the job depends on the assumed substitution elasticity between college and high-school graduates). However, the (nearly) stable demand shift masks the fact that for college graduates the between-industry shift in demand slowed down in the 1980s while the within-industry shifts accelerated from 1967 to 1987.

In our supply-demand-institutions framework several other potential explanations are to be considered. In the following we briefly review explanations that are not extensively discussed in the literature.

As we have seen, *measured* labour supply shifts could only partially explain the observed trends. It is possible, though, that *unmeasured* supply changes are important. The inflow of illegal immigrants is one of these unregistered supply changes (this inflow is probably an unmeasured increase in low-skilled labour). Leamer (1994) and Bound and Johnson (1992), among others, suggest this explanation. It is beyond the scope of this chapter to deal extensively with this issue (see for example the book edited by Borjas and Freeman, 1992).[9] Alternatively, a decline in educational quality of the low-educated workers is also an implicit supply change (that would also result in an increase in the education premium).[10] Opinions about the importance of these unmeasured supply changes are rather diverse.

We have already stated that demand changes, besides supply changes, are an important part of the explanation. The only element of the framework, not yet discussed, is the institutional environment. Here several changes occurred that might have been relevant. First, declining unionization is observed. Bound and Johnson (1992) test directly for the influence of unionization and indirectly via the importance of rents and conclude that de-unionization does not explain a very large part of the relative wage change. Second, decreasing relevance of the minimum wage might play a role; only a small proportion of the workforce can be affected by this institutional change and besides that, the increased wage dispersion is also caused by changes in the upper part of the wage distribution. Third, deregulation might have affected wage setting behaviour. One can think of a lot more government measures, like changes in the social security system, that might have affected the wage structure somehow.

For several reasons we do not explore these (potentially relevant) institutional changes further: (i) the ones that are documented in the literature have played a minor role, (ii) there is no reason to assume that these measures did coincide internationally, as is required given the facts documented above.

The results reported above do not provide a comprehensive explanation for the observed changes. They only point out that demand changes are necessary to provide an explanation. Thus we need to explore possible explanations for a demand shift.

4. TRADE: AN EVALUATION ON THE BASIS OF THE 2 x 2 x 2 HECKSCHER-OHLIN MODEL

The observed increase in the gap between the factor rewards for high- and low-skilled workers, together with the more pronounced presence of low-wage countries in the global economy, led many observers to conclude that the latter caused the former. The argument that global market integration led to the increase in the wage gap is fairly persuasive for many people. Economists have also been sympathetic to the argument, as there is a theorem in international trade theory, named after W. Stolper and P. Samuelson, that establishes a clear relation between global markets and internal factor rewards. The concept is clear but the relevance of this concept for the issue at hand is strongly debated by several authors. In this section the logic of the Stolper-Samuelson theorem is set forth, and several tests of the relevance of the Stolper-Samuelson theorem are reviewed.

4.1 Theory: The 2 x 2 x 2 Heckscher-Ohlin model

Stolper and Samuelson (1941) established one of the theorems in the endowments-based international trade theory connected with the names of Heckscher and Ohlin. To illustrate the logic of their theorem[11] we will run through an example based on the situation in the US. Suppose the US is skilled-labour abundant. Furthermore we assume that the US starts trading with LDCs due to a decline in trade barriers erected by the LDC. The US will shift production towards the skilled-intensive goods industries (the relative price of the skilled-intensive good is increased). As the shrinking sector is unskilled-intensive, it releases relatively plentiful unskilled workers. Thus, the relative wage of the unskilled has to decline in order to get the unskilled absorbed in the skilled-intensive industry. This change in the relative factor reward changes the factor-input ratio in the both industries. Summarizing the results after the market has settled at the new equilibrium:

1. unskilled labour in the US will be harmed relatively and absolutely,
2. the ratio of skilled to unskilled in the US declines in both industries,
3. employment is shifted towards the skilled-intensive industry in the US,
4. the US exports the skilled-intensive good,
5. the price of the skilled-intensive good rises relative to the unskilled-intensive good and,
6. for the LDC, implication (1) to (5) are reversed.

Implication (1) is the basic observation that led to the Stolper-Samuelson theorem.[12] In the remainder of this section the literature that directly or indirectly analyses the predictions (2) to (6) is reviewed. Box 2.1 provides a graphical exposition of the Stolper-Samuelson theorem.

BOX 2.1 THE STOLPER-SAMUELSON THEOREM IN THE LERNER-PEARCE DIAGRAM

To illustrate the Stolper-Samuelson theorem the Lerner-Pearce diagram is employed. This graphical technique highlights the predictions of interest for the wage-inequality debate. This box explains the Lerner-Pearce technique and explores the Stolper-Samuelson theorem.

Preliminaries

The axes in Figure B2.1 denote factor inputs in production; skilled (H) and unskilled (L) workers required for the production of the skill-intensive good S and the unskilled-intensive good U. The curves denote the combinations of factors of production required to produce the production volume of goods S and U that generates a value of unity. To calculate the value, the production volume is multiplied by the given world market prices. Hence,

$$SP_S = 1 = w_L L + w_H H, \quad UP_U = 1 = w_L L + w_H H. \tag{2.1}$$

From these relations it follows that for both equations:

$$\frac{dH}{dL} = -\frac{w_L}{w_H}. \tag{2.2}$$

The equilibrium factor-price ratio can be determined by the angle of the tangent with the horizontal axis. Obviously, good S is produced relatively skilled-intensively as indicated by the rays from the origin.

The Stolper-Samuelson theorem

Now suppose the price of the unskilled-intensive good (P_U) falls. Production has to increase to produce a unit value, hence the U-curve shifts outward to U'. The new w_L/w_H is lower than previously and both goods are produced with a more unskilled-intensive technique.

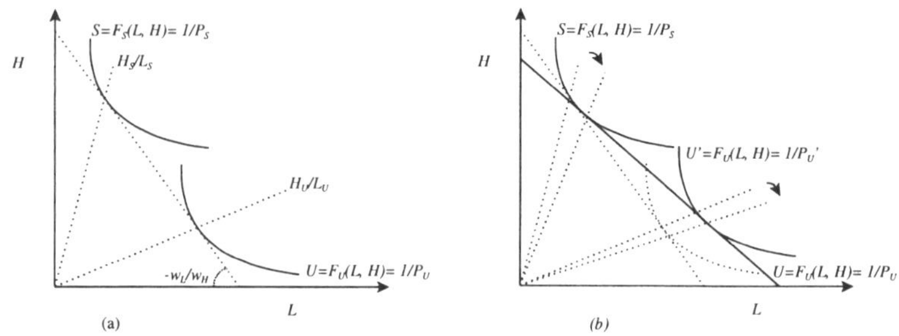

Figure B2.1 The Stolper-Samuelson theorem

4.2 The ratio of skilled to unskilled workers declines

The Stolper-Samuelson theorem predicts that within all industries in the US, the ratio of skilled to unskilled workers should decline. The method to examine this is the following. First, one needs to distinguish skilled and unskilled workers in the labour force. Second, the relative wage increase of skilled workers should be met with a decrease in the relative employment of the skilled workers.

Krugman and Lawrence (1993) perform some analysis on this implication. Skilled versus unskilled workers are approximated by blue-collar and white-collar workers. The change in the ratio between these blue- and white-collar workers is analysed for the manufacturing sector.[13] The manufacturing sector is decomposed in industries at the 2-digit SIC level. Opposite to the prediction, Krugman et al. find for nearly all of the observations an increase in the ratio of skilled to unskilled workers within industries.

Lawrence and Slaughter (1993) perform a similar test. They distinguish skilled and unskilled workers by respectively non-production and production workers. At 2, 3 and 4 digit industry level they find that the ratio of skilled to unskilled workers increases; again opposite to the theoretical prediction. Their conclusion is that the Stolper-Samuelson effect is (at least) overwhelmed by something else.[14]

With respect to this test of the Stolper-Samuelson theorem there are two vital theoretical limitations. The first is that the logic of the Stolper-Samuelson prediction of an increase in the ratio of low- to high-skilled workers holds only in the case of *unchanged endowments*. Three influences that might violate this condition are the following. First, the decrease in the size of the manufacturing sector should be taken into account. Second, the relative supply of high-skilled workers increased substantially during the 1980s.[15] Third, changes in unemployment might also affect endowments.[16]

A decreasing size of the manufacturing sector provides the connection with the second issue of why *the increase in skill intensity is a disputable basis for rejecting the role for trade*. In Section 6 we will stress more extensively the following: the assumption of homogenous and unchanging production functions within industries and equal production technologies across borders is unrealistic and disturbing if one wants to explore the role of trade. We want to distinguish more activities within one single statistical industry, but more on this later.

Besides the theoretical limitations, there is a serious empirical limitation. This limitation concerns the appropriateness of the production/non-production workers division to approximate unskilled and skilled workers. Leamer (1994) stresses clearly that this approximation is troublesome. Workers occupied with product development are for example classified as production workers, whereas workers performing routine office functions and those delivering sales are categorized in the non-production worker category. Similar problems exist with the blue/white-collar distinction.[17]

4.3 Employment shifts towards the skilled-intensive industries

An employment shift toward the skilled-intensive industries should be accompanying the relative wage trend if the Stolper-Samuelson theorem tells the true story. A test for this prediction is simple: order industries by skill intensity and analyse whether the employment growth is most pronounced for the skilled-intensive industries.

Krugman and Lawrence (1993) test for this prediction. Comparing the relative skill intensity in 1979 together with the growth of employment between 1979 and 1989 for the industries they distinguish, reveals that the skill-intensive industries showed no tendency to grow faster.

The limitations to this test are essentially similar to those that apply to the first test. Theoretically, we will argue that more activities should be distinguished within industries (or, equivalently, that the level of disaggregation is too limited). Therefore, denoting the skill intensity of an industry, consisting of a rather broad set of production activities, as a single number is precarious.

The previous remark might also be interpreted as an empirical limitation, but besides this one, the used identification of a worker's skill level remains of course uncomfortable.

4.4 The US exports the skilled-intensive good and imports the unskilled-intensive good

Economists with limited knowledge of the Stolper-Samuelson theorem have tried unsuccessfully to estimate or assess the impact of trade on the domestic economy using the volume of trade as an indicator. This opinion is most forcefully expressed by Leamer (1994, 1995, 1996b). He speculates that it is probably the Stolper-Samuelson prediction that the US imports the unskilled-intensive good that induces the factor content of trade calculations. The argument for the use of the so called Factor Content of Trade (FCT) calculations is appealing; one estimates what *would* have been the demand for factors of production *had* the domestic economy been isolated from world trade.

The factor content of trade methodology is rather familiar, and therefore the explanation – based on Wood (1994) – will be brief. Factor exports and factor imports respectively are:

$$z_{x,US} = A_{US} x_{US}, \qquad z_{m,US} = A_{US} x_{LDC}, \tag{2.3}$$

where x_{US} is the vector of sectoral shares of the manufacturing exports, A_{US} is the factor use matrix and thus $z_{x,US}$ is the factor content coefficient for exports. A similar vector can be calculated for imports ($z_{m,US}$). If we now assume that US exports are balanced with LDC exports, we are able to calculate the change in factor demand had the US produced the goods themselves by:

$$Z_{US} = X_{LDC}(z_{x,US} - z_{m,US}), \tag{2.4}$$

where X_{LDC} is the vector of net import values.

Katz and Murphy (1992), discussed in Section 3, analysed the underlying causes of the increased education premium and found ‘some’ role for trade. Katz and Murphy estimated the impact of trade using the FCT method and found that the direction of the trade-induced shifts was consistent with the observed pattern, but that the magnitude was (too) small.[18]

Sachs and Shatz (1994) vary slightly on the FCT approach by calculating[19] counterfactual employment shares. Via an input–output matrix, final demand for a sector is estimated for 1990 keeping the import-penetration quote at the level of 1978. While being aware that the following procedure is a serious shortcut, the counterfactual output levels are *directly* translated into factor demand. Next they order industries by skill intensity and identify the relevant trade partners for these industries (DC or LDC). They find that less-skilled-intensive sectors are hit relatively hard and that the resulting demand shift of 5.1 per cent toward non-production workers is primarily caused by trade with LDCs. The conclusion that a tendency towards increased wage differentials *might* arise from this employment shift is the only one possible.

FCT calculations are based on the magnitude, or volume, of trade. Some authors use the volume of trade directly to underpin their arguments: Krugman and Lawrence (1993) support their conclusion that it is implausible that trade played a substantial role in widening the wage gap by arguing that the bulk of US trade is with high-wage countries and that trade in the US is still a small fraction of output. In later work Lawrence (1994) reiterates this as follows: 'US imports from developing countries did increase rapidly over the decade, but again what needs to be born in mind is the magnitude.' (p. 13). Thus their argument is basically that the magnitude of trade with low-wage countries is simply too small to affect wages substantially.

Murphy and Welch (1992) also proxied the impact of trade by a volume indicator: the net import of durables as a share of GNP. Trade is found to be a significant determinant of wages. Murphy and Welch explicitly state that they are not convinced that trade is the ultimate cause for the observed changes; biased technological progress might also be relevant.[20]

The limitations of the FCT approach have led to a heated debate in itself. Leamer condemned work based on FCT as: 'incompetent, immaterial and irrelevant' where Krugman (1995b) called FCT: 'a procedure that turns out to be entirely reasonable' (p. 38). Leamer's (1994, 1995, 1996b) critical argument runs as follows. The Stolper-Samuelson theorem connects factor rewards with goods prices via zero profit conditions, without the volume of trade entering the formulation of the theorem. To clarify: suppose a country faces increased competition in unskilled-intensive goods, but that it succeeds in keeping a constant market share by lowering the wages of unskilled employees. When asking the question whether trade can be blamed for this observation by means of calculating the change in the FCT, the answer will be no. This is obviously the wrong answer as ex-post trade volumes are an unreliable indicator of the ex-ante threat.[21] To restate the argument briefly: factor prices are set on the margin.

In a clarifying paper, Deardorff and Staiger (1988) show that the translation from FCT to domestic factor prices is indeed not as obvious as it seems. By constructing an 'equivalent autarky equilibrium', that is by adjusting domestic resources for the amount resources that trade contends, they are able to exclude only the counter-intuitive result that factor prices move along with endowments under fairly restrictive assumptions (unitary substitution elasticity on the demand and production side). Leamer (1996b) reiterates and extends the analysis by Deardorff and Staiger and adds to their conclusions that only the repercussions for earning *shares* (not levels) of a complete elimination of trade can be established under the restrictive assumptions mentioned. More intuitively: the problem is that factors embodied in trade are determined simultaneously by tastes, technologies, factor supplies, external goods markets and the deficit/surplus on the trade balance.[22] Without being exhaustive, Leamer (1996b), Bhagwati and Dehejia (1994), Sachs and Shatz (1994) and Lawrence and Slaughter (1993) conclude that it is the development in goods prices that should be analysed.[23]

Krugman (1995b) calls the FCT calculations an 'entirely reasonable' exercise, with the restriction that trade deficits should be handled with care. How is this radically different view possible? First of all, Krugman is eager to accept the restrictions following from the Deardorff and Staiger analysis as being not too far from realistic. The second and more fundamental difference is Krugman's rejection of the argument that prices can be treated as exogenous. Throughout the debate this issue causes a watershed. The proponents of the price approach consider a country as price taker, the others see relative world market prices as endogenous. Krugman's (1995b) argument runs as follows. First, he asserts that the US, let alone the whole OECD, will alter goods prices if technology or factor supplies are changed. Thus, the price approach is incapable of disentangling domestic and foreign causes. As we are interested to know what wages would have been had *only* the trade from low-wage countries been impossible, the volume of trade is a crucial piece of evidence.[24]

So we are left with two compelling arguments favouring two different approaches. The verdict on the appropriate methodology (and with it the usefulness of FCT) depends on whether or not prices can be treated as exogenous. Whatever the verdict, however, one empirical problem remains and that is how to treat the trade deficit.

4.5 The relative price of the unskilled-intensive good declines

The shock that drives all the observations following from the Stolper-Samuelson theorem is that the price of the unskilled-intensive good declines. This is translated, via the zero-profit conditions, in factor rewards. A lower relative factor reward for the low skilled in the relatively unskilled-scarce country results. Prices are a reliable indicator as it is the beginning of the chain of causation.[25] A test based on prices of unskilled- versus skilled-intensive goods has to adjust for differences in Hicks-neutral technological change.

In search of the answer to whether or not foreign competition plays a role, Lawrence and Slaughter (1993) identify the beginning of the chain of causation. They test whether prices of unskilled-intensive goods decline relative to skilled-intensive goods. This appears not to be the case; they find the opposite. No more is necessary to show that trade is not the main cause for the observed increase in wage inequality.

Sachs and Shatz (1994) also look at relative price changes. Contrary to Lawrence and Slaughter, the view that the Stolper-Samuelson mechanism was at work is supported. The notion that in the calculations in both papers the production/non-production worker ratio is used, leaves differences in the data as the obvious explanation for the opposite result. Sachs and Shatz indeed point to the different use of data. They add a dummy for computers. This is essential for their result. They admit that what they find is not enough to explain the huge increase in earnings inequality and point to the interaction between asymmetric technological development and the lowering of low-skilled-intensive goods prices.

Leamer (1996a) performs a somewhat different exercise over a more extended period, namely from 1970 to 1990. Leamer starts with 'the fundamental condition linking product price changes, factor cost changes and technology changes' (p. 23):

$$\hat{p} = \theta\hat{w} - T\hat{F}P \ , \tag{2.5}$$

where p denotes product prices, w factor prices and TFP technology. Leamer tries to disentangle the part of factor-price changes that is due to technological change and the part that is due to globalization (θ are factor shares). This equation should be interpreted as evidence that goods-price changes lead to factor-price changes and that productivity increases lead to factor-price increases *if* prices are fully determined on the world market. The theory, however, says nothing about the relation between total factor productivity (TFP) and price changes. An assumption on how much of the productivity gain is passed on to lower prices should be made to be able to calculate the technological effect on wages. Leamer takes two extremes: full pass-through and no pass through. If the sector-specific TFP change is fully passed on into lower prices, there is no room for (sector) biased technological change to affect factor rewards (the only technology effect on factor rewards is the average TFP change). Given the pass-through assumption, in the equation denoted as λ, the wage change attributable to technological change is:

$$(1 - \lambda)T\hat{F}P = \theta_i \hat{w}_{tech} \ . \qquad (2.6)$$

The remainder[26] is due to globalization:

$$\hat{p} + \lambda T\hat{F}P = \theta \hat{w}_{glob} \ . \qquad (2.7)$$

The regression analysis is based on the latter two equations. The left hand side – sectoral TFP growth and sectoral inflation corrected for TFP induced changes – is the dependent variable. Whereas the (beginning of the period) factor shares are the independent variables. The estimated coefficients are mandated factor price changes, that is, changes that are consistent with no changes in profits. In the 1960s the price/globalization effect worked strongly against capital, but the technology effect compensated fully. In the 1970s the globalization effect worked strongly against low-wage workers and the technology effect worked weakly against the low-wage workers. In the 1980s the globalization effect worked out strongly against capital and favoured high- and low-wage labour, where technology affected high-wage workers unfavourably.

Thus the 1970s is the period where the action is with respect to globalization. Two remarks need to be made. First, the early timing of the globalization effect driving down the rewards for low-wage workers is to some extent explained by pointing at lags in spreading this effect in the economy. Leamer, however, conceives the timing of the wage-gap increase in the 1980s as inaccurate.[27]

Krueger (1997) extensively discusses what he calls, the 'price puzzle'. Using a new data set, the relation between skill intensity and price development is estimated once again. The advantages of the Krueger data[28] are a 4-digit level of aggregation and a more recent period (1989–95). Krueger convincingly reconfirms what Sachs and Shatz (1994) find: a positive relation between skill intensity and price growth that is robust for several indicators for skill intensity.

A theoretical limitation in (early) work on relative prices is pointed at by Davis and Topel (1993) in commenting on the work of Lawrence and Slaughter (1993) on the relative prices: 'Once technical change is accepted as a major driving force behind relative wage movements, relative price movements for traded goods no longer provide an accurate indicator of the direction or magnitude of trade's effect on relative wages.'(p. 220). This comment cuts ice, as Lawrence and Slaughter in their analysis of sector bias in technological change assume given prices. Thus it is implicitly assumed that technological change is a local phenomenon. Krueger (1997) does not use TFP data to correct relative price changes either. He argues that TFP changes will harmlessly add to the error term if TFP growth is uncorrelated with the other variables explaining price growth. Sachs and Shatz (1994) and Leamer (1996a) on the other hand adjusted their prices for TFP changes before assessing the Stolper-Samuelson price changes.

A clear empirical limitation of the test for this Heckscher-Ohlin prediction is that measuring trade prices is notoriously difficult. For example, because of changes in the product mix and quality of the goods. Correcting for TFP changes is an exercise that is known to be inherently problematic also.

Therefore it might not be so surprising that we end up with conflicting results in different papers.

4.6 For the low-skilled-abundant trade partners the predictions are inverted

As is assumed, formerly erected trade barriers between LDCs and the US are considerably reduced. The Heckscher-Ohlin model predicts for the low-skilled abundant LDC exactly the opposite movements in production and prices to what is predicted for the US. As studies for the low-skilled abundant regions are scarce and data availability is limited the summary of results is restricted to the prediction for factor rewards. The gap between low-skilled and high-skilled worker wages should decline.

Wood (1994) finds for southern countries a reduction in the wage inequality, as predicted by the Heckscher-Ohlin theory. As Wood distinguishes three types of labour, the prediction of declining overall inequality is not obvious. Lowering trade barriers causes the south to specialize towards basic-educated-intensive goods. The wage differential between high skilled and those with basic education should decline, but the differential between the average high-skilled/basic educated and those without education should widen. Thus the prediction with respect to overall inequality is ambiguous, but Wood expects a decline in inequality and finds it.

A cross-country regression of inequality and export shares in manufacturing yields a significant negative sign, but it is probably not too hard to convince anyone that this is not very informative. Time series analyses yield again the expected sign but the results are hardly significant. Besides, in such a time series analysis one should account for supply changes to be able to distil a role for trade with the north.

Next the changes in Mexican (and Moroccan[29]) relative wages are discussed. Why discuss Mexico separately? First, the Mexican changes in the 1980s have some resemblance to a laboratory experiment.[30] Second, the Mexican economy is well documented in the literature. Major changes were a considerable liberalization of trade and the announcement of Mexico to join the GATT. Hanson and Harrison (1995) test whether observed changes in relative wages fit in the Heckscher-Ohlin framework. As the US is Mexico's most important trade partner[31] and the US is relatively skilled-labour rich, the Heckscher-Ohlin prediction is a declining inequality between skilled and unskilled workers in Mexico. The reality differs though. Between 1984 and 1990 the relative wage of the skilled rose by 33 per cent. Hanson and Harrison took this as a starting point to make up stories, or call it theories,[32] that were consistent with Heckscher-Ohlin predictions *and* the observed wage-inequality pattern. After testing for the implications of these theories it is concluded that there is no way that the facts can be reconciled with the Heckscher-Ohlin theory.

Even those who are sympathetic to the explanation that the increase in the wage gap is due to non-domestic factors, that is trade, admit that the evidence with respect to the 2 x 2 x 2 Heckscher-Ohlin model is at least mixed. The ratio of skilled to unskilled workers increases instead of decreases at all levels of disaggregation that have been considered. Employment did not clearly shift in the direction of skilled-intensive industries. If one accepts the FCT calculations as valid and informative, the magnitude of Heckscher-Ohlin tendencies is too small. The evidence of the evolution of relative prices, the most direct measure of the Stolper-Samuelson effect at the beginning of the chain of causation, is mixed. Finally, the relative wage change in Mexico and Morocco is clearly opposite to the direction predicted by the Heckscher-Ohlin model.

Thus, irrespective of the discussions on the appropriateness of the approach and on the limitations of some tests of the Stolper-Samuelson predictions, it is fair to state that the evidence is mixed. To formulate briefly what is at stake we use the evidence that *is* clear so far, to formulate the 'trade and wage puzzle': 'The relative wage increase of skilled workers is accompanied by an increase in the ratio of skilled to unskilled workers within industries.'

How to plough ahead given these findings? The fact that the predictions belonging to the 2 x 2 x 2 Heckscher-Ohlin model are not confirmed basically led to two approaches. One approach is to look for domestic causes, as the above reported evidence suggests that non-domestic causes are unlikely to be important. The search for a domestic cause, some form of biased technological change, is subject of the next section. The other approach is not to reject a role for trade, but to refine or reformulate the Heckscher-Ohlin model. Section 6 deals with this.

5. BIASED TECHNOLOGICAL CHANGE: AN EVALUATION OF DOMESTIC CAUSES

The weak support for the trade-related explanation led authors to the alternative explanation of biased technological change. The logic is simple: the relative supply changes did not contribute much to the explanation of inequality, clear evidence for a change in demand due to trade is not found, thus there is left a residual that can only be attributed to technological change. This explanation of biased technological change is prompted by the common-sense idea that new technology mostly replaces unskilled workers; for example one worker operates the robot-system that replaced several assembly-line workers. This would, however, imply that the productivity of the unskilled workers goes up. This section sorts out what type of technological change solves the 'trade and wage puzzle'.

5.1 Theory: The 2 x 2 model

To analyse what is meant by biased or asymmetric technological progress, again the two sector, two factor model for the small-open economy is used. We assume that technology is locally available.

The skill-biased technological change explanation comes essentially in two different forms: a sector bias in technology or a factor bias in technology.

A sector bias

Take a Hicks-neutral sector bias in technological change. Hicks neutrality implies that at the original factor-price ratio the factor intensity is unaffected (see Figure 2.1a where the lines of tangency indicate the factor-price ratio). Figure 2.1b depicts the new equilibrium factor-price ratio after a relative increase in the productivity of the skill-intensive good. Increased inequality (a decrease in w_L/w_H) results and hence the first part of the 'trade and wage puzzle' is solved. The skill intensity of production (depicted by the rays from the origin), however, decreases. Thus the second part of the puzzle is not solved.

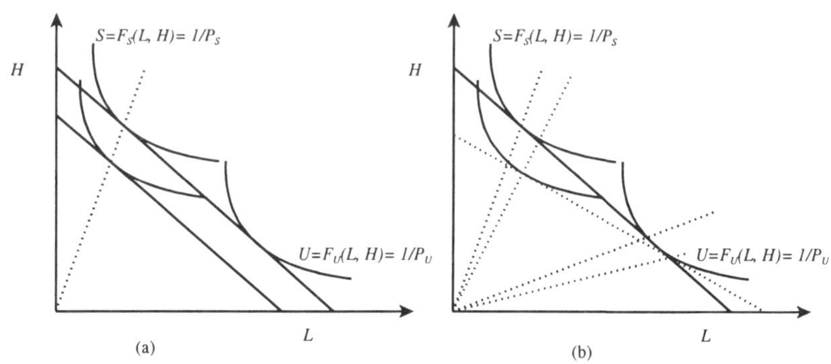

Figure 2.1 *Sector-biased technological change*

A factor bias

Skill-biased technological change is a change in the unit factor requirements at given factor prices. In the Lerner-Pearce diagram this is depicted by a shift in the unit-value curves along the factor-price ratio line (see Figure 2.2a). One cannot read it directly from the figure but it is easily inferred that to maintain domestic factor-market equilibrium, production should shift towards the unskilled-intensive good. Figure 2.2b shows a more plausible pattern; skill-biased technological change that also reduces costs (in a sector-neutral way).

Evaluating the results with respect to the formulated 'trade and wage puzzle' reveals that the skill intensity of production increased but that the factor-price ratio is unaffected. It is the observation that a sector-neutral skill bias by definition does not affect relative wages that causes Leamer (1995, p. 5) to argue that, 'factor bias doesn't matter, sector bias does'.

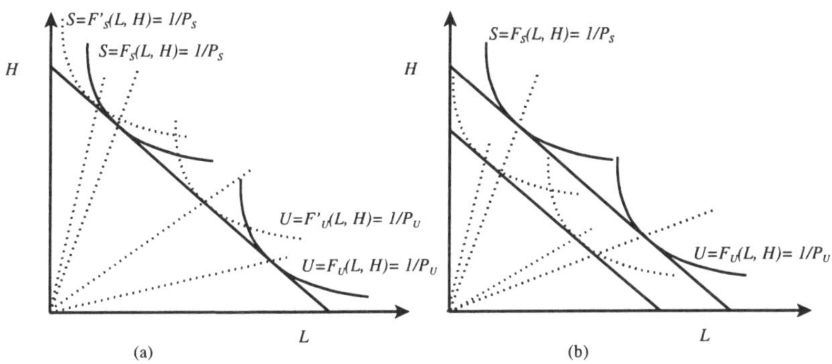

Figure 2.2 A factor bias in technological change

Leave the small-open-economy perspective for a moment and assume that the economy is sufficiently large to affect the world market price. The change in production structure necessary to keep production factors fully employed (increasing the production of the unskilled-intensive good) is likely to exert downward pressure on the price of the unskilled-intensive good (see curve U' and note that $P_u' < P_u$. Figure 2.3 shows that relative wage deterioration of the unskilled workers results (along the increased skill-intensity of production). Hence, this perspective solves the trade and wage puzzle.

The discussion so far yields two lessons. First, it is likely that skilled- instead of unskilled-saving technology is the principle cause for inequality. This is easily seen when it is recognized that unskilled-labour saving technology is the equivalent of increased productivity of the unskilled. Second, the 'trade and wage puzzle' (the combination of increased relative wages for skilled workers with the increased skill intensity of production) is 'solved' by biased technological change in two ways. One possibility is that in the large open economy a skill-bias leads to a decrease in the price of the unskilled-intensive good. Or, equivalently, a combination of two shocks on the small open economy: skill-biased technological change together with a decrease in the relative price of unskilled-intensive goods (due to trade liberalization for example). A second possibility holds for the closed economy: a skill-biased technological improvement in combination with ample substitutability between both goods.

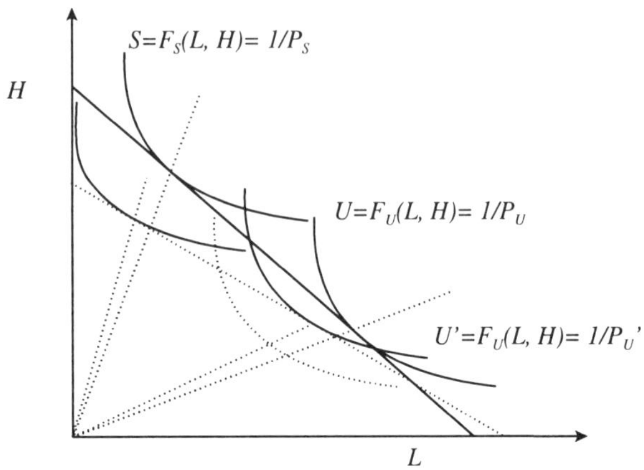

Figure 2.3 A large open economy

5.2 Theory: Endogenous technological change

We are looking for the basic behavioural assumptions or structural changes that caused an increased skill premium. The following notions are preliminary and there is hardly any evidence supporting or rejecting these speculations.

That necessity is the mother of invention is something hardly disputable after a decennium of endogenous growth theory. If it would be asymmetric technological progress that drives the observed trend, what is the incentive for developing such a technology?

First, the incentives for generating a sector bias by endogenous technological change are considered. Two directions can be followed. The first is that increased competition lowers the future stream of profits and therefore the potential gains from investment in R&D. When we assume that competition increased mostly in unskilled-intensive sectors causing these sectors to perform less R&D (what ultimately causes a sluggish TFP development) this might lead to trade-induced bias in technological change. This is biased in the sense that it differs across industries, according to skill intensity (see Figure 2.1).

A second line of reasoning works in a similar direction. Assume wages of skilled workers tend to rise relative to unskilled workers' wages (due[33] to trade?): as R&D activity is thought to be skilled-intensive, the relative wage trend makes producing more attractive relative to innovating, at least in industries where production is low skilled-intensive. Thus this mechanism might again work out asymmetric across industries, where again low skilled-intensive industries face lower incentives to innovate.[34]

Third, the increase in the relative endowment of skilled workers could induce a bias in endogenously generated technological change by one of the following mechanisms. First, the market size for innovations that improve the productivity of skilled workers increases with the number of skilled workers. Hence, an increased endowment of skilled workers, as such exerting downward pressure on their relative wage, might induce complementary innovations that cause the relative wage of skilled workers to increase in the long run. This implies that the demand curve for skill is upward sloping (see Acemoglu, 1998). Second, an increase in the relative endowment of skilled workers implies that the knowledge stock is more valuable now and in the future, as skilled workers work with the knowledge stock. This can again result in an upward-sloping demand curve for skill (see Chapter 6).

Note that the first two reasonings on endogenous technological change both hinge on trade or at least are likely to be related to trade. The last argument is easily developed within a closed economy framework.

5.3 Evidence

A fair amount of 'evidence' in favour of the biased-technological-change explanation consists of rejections of trade-related explanations. The approach of rejecting trade as an explanation and assigning the unexplained residual to technology is common practice. There are in principle no objections to a proper application of this strategy. The keyword here is 'proper', however; a discussion on whether this strategy has been applied accurately is postponed. In this section *direct*[35] evidence for 'biased technological change' as an answer to our basic question will be reviewed.

Using micro data, Krueger (1993) provides direct evidence on the *type* of technological change that is relevant. Using CPS data for 1984 and 1989, the impact of the 'computer revolution' is assessed. As a take off some descriptive data: computer use increased by 50 per cent, to 38 per cent of the workers in the sample. The use increases strongly by educational attainment and the use is very widespread in sheltered industries as insurance, banking and public administration. In a Mincerian wage equation, a dummy for computers is added. In 1984 the use of computers yields an additional pay-off of 18.5 per cent, in 1989 this is slightly higher.[36]

One serious limitation, as always, remains: is it actually the use of computers that is valued by employers, or does the variable represent something else that is omitted in the regression?[37] Taking the results for granted – for a minute – Krueger explains half of the increase in education premium, and even more when the interaction between computer use and education is taken into account. DiNardo and Pischke (1997) first replicate Krueger's result using a German data set. Next, they show that computer use reflects some unobserved heterogeneity in the data set. Sitting down while working and using pencils also has, alongside computer use, a significant effect on wages. These findings cast serious doubt on the interpretation of the computer-use premium as reflecting a productivity bias.

In view of the previous discussion on the theory, Krueger's work might lead us to accept the view, opposite to the popular one, that it is indeed skilled-labour-saving technology that is relevant. But remember that in the open economy case the skill intensity within sectors moves in theory in the opposite direction from the one actually observed (see Figure 2.1).

Simultaneously with suggesting that asymmetric technological progress is the plausible explanation for the observed trends, several authors refer to

Bartel and Lichtenberg (1987) and Mincer (1991) for supporting evidence. Bartel and Lichtenberg did pioneering work on capital/technology-skill complementarity. They established a relation between skill composition of the workforce and the capital stock. Industries with newer capital or higher R&D intensity have a more educated workforce.

The capital/technology-skill complementarity, found in earlier cross-sectional studies, establishes no causal relation. Estimating this in a time series framework would be an improvement. Mincer (1991)[38] relates in a time-series approach the return to human capital to the pace of technology correcting for other factors (such as, for example, supply changes). R&D indices do rather well as a proxy for technology.[39]

Berman, Bound and Griliches (1994) report findings for the manufacturing sector.[40] The question they want to answer is what causes the increase in the fraction of non-production workers (or their part in the wage bill) in manufacturing? There is no clear underlying model they use. Between 1979 and 1989, the number of production workers declined by 15 per cent whereas the number of non-production workers increased by 3 per cent. Between and within industry shifts are distinguished; the latter explains one third, the former the remaining part. Their first contribution is to subdivide the between and within shifts in end-use categories: exports minus imports, domestic consumption and defence.[41] Berman, Bound and Griliches conclude that trade cannot be a principal cause of the observed trends in the wage bill. Biased technological change is offered as an alternative. Their second contribution is that they try to measure directly the importance of biased technological progress. A cross-sectional regression is used in the following manner. The annual change in the non-production workers' share in the wage bill is the dependent variable. Independent variables are time dummies and the change in capital intensity. The data on other production factors allow complementarities to be identified. A capital/skill complementarity,[42] suggested by Bartel and Lichtenberg (1987), is confirmed. Next, they add the importance of computers and R&D expenditures over sales as independent variables. They contribute significantly to the explanation of the increase of non-production workers' wage bill share. Berman et al. conclude that asymmetric technological progress is the major cause for the shift towards non-production workers; trade might have been a minor amplifying effect.

We already encountered the work of Lawrence and Slaughter (1993) in the assessment of the 2 x 2 x 2 Heckscher-Ohlin model. Having rejected this model, technological change is left as an explanation. First, they test whether Hicks-neutral changes/differences in technology between industries are relevant. They assume given prices. Two predictions are to be tested (see Figure 2.1). The ratio of skilled over unskilled workers should fall. As we already know, it does not. A more direct prediction to test is the notion that total factor productivity growth should be most pronounced in skilled-labour intensive industries. This sector bias in technological change is (not convincingly) found. For Lawrence and Slaughter, nothing else is left than to point at factor-biased technological change as a possible explanation.

Leamer (1994, 1996a) comments on methodology and presents also some preliminary empirical work. Leamer finds[43] (as reported on in Section 4) that technology clearly has decreased inequality in the 1980s.

In a promising approach, Baldwin and Cain (1997) assess the dominant factor causing relative wage changes. In the context of a general equilibrium model, technological change, relative endowment shifts and alterations in the trade pattern are simultaneously taken into account.[44] Exceptionally, they use a large open economy framework. Using various data sources they assess which of the three driving forces is consistent with the observed pattern in goods and factor prices, skill intensities and output changes. For the late 1960s to the early 1970s the relative supply increase of high-skilled workers is important. For the 1970s the relative supply increase together with an unfavourable bias against unskilled workers is the driving force. And for the 1980s their analysis points again to biased technological change.

Baldwin and Cain (1997) point to the theoretical limitation that the approach is not capable of discriminating between all potential explanations. A second limitation of a theoretical nature is common in the work on this topic; the problematic distinction between high- and low-skilled workers.[45]

Besides the usual limitations to econometric work there is a more fundamental problem. In the studies discussed in this section, industry data are used in regression equations to explain wage trends. The essential difficulty is that industries are treated as homogeneous entities. It is hard to argue that industries are representing a single activity. What is meant is that industries consist of several activities and that activities can be distinguished and ordered by 'technology intensity' and skill intensity. Taking this conjecture for granted the above cited regressions are not as informative as they seem at first sight. Suppose that trade causes outsourcing or a shut down of the less skilled-intensive parts within industries. It is likely the less skill-intensive parts are also 'technology extensive'. With the combination of these two assertions the conclusions connected to the regressions could also be a consequence of trade-induced outsourcing.

The argument that highly disaggregated data are used cuts no ice. The fact that a 4-digit level disaggregation is not sufficient is a problem that Berman et al. recognize but surprisingly further ignore: '*Even within specific industries* this "foreign outsourcing" is likely to have disproportionate effect on less skilled labour...' (p. 376, italics added). More on this issue in Section 6.

An exceptional approach is followed by Bound and Johnson (1992)[46] when they evaluate alternative explanations for the changes in the structure of wages. Bound and Johnson try to answer the question on the causes of the observed changes in the wage structure by estimating the contribution of alternative explanations in the context of a clear theoretical framework for the *closed* economy. The framework allows for some autonomous demand shift (caused by foreign competition or a preference shift). Criticizing this theoretical framework is easy,[47] but it is their merit that the framework is clear and that all potential explanations fit in. Explanations are divided into (a) changes in the non-competitive part of wages (the rent) and the impact of workers changing between industries with different rents (see Krueger and Summers, 1988), and (b) in changes in the competitive part of wages. The latter is subdivided into changes caused by supply, demand, and technology. Demand shifts over 11 industries have a modest (sometimes even 'wrongly' signed) impact on relative wage changes. The change in factor-specific technology contributes most importantly to the changes in relative wages;

after accounting for this, hardly anything remains unexplained. This is not surprising, as they state: 'The major difficulty with this explanation, unlike the explanations involving industry wage effects, supply and product demand, is (as in the analysis of the sources of economic growth) that it involves the residuals of the intra- factor demand function rather than directly observable phenomena' (p. 383). Besides skilled-labour-biased technological change they think changing unmeasured labour quality is a principal reason for the increased wage differential.

6. AMENDING THE TRADE MODEL

The weak performance of the basic trade model, reported on in Section 4, guided research in two directions. The first direction is the analysis of technological change favouring skilled workers. The rationale for pointing at biased technological change was that the other explanations did not contribute substantially to explain the observed wage trend. To speak with Leamer, their reasoning was: 'Eliminate the impossible, and all that is left is truth' (1996a, p. 2). In this section we will argue that on the basis of the existing literature it has *not* been shown convincingly that all the other explanations are eliminated. This line of reasoning induces a second strand of continuing research. In this section we extensively discuss approaches related to international trade, but deviating from the basic 2x2x2 Heckscher-Ohlin model. The logic that lacking evidence in favour of the basic trade model does not imply a negligible role for trade is blunt, but often 'forgotten'.

6.1 Disaggregation, non-homogeneous goods and outsourcing: A new assumption

A fundamental change in an assumption behind the basic trade model is proposed in this section.[48] The assumption that is altered in this section is that goods produced in one industry are homogeneous. The hypothesis postulated here is that it is possible to distinguish more activities within a single industry. First, the rationale for this deviating hypothesis will be developed. After that, an example of a theoretical implementation of this assumption is discussed together with estimates for the impact of trade on wages.

To understand the rationale for the deviating assumption it might be clarifying to take a closer look at industries. To set the stage, how do indigenous firms respond to increased competition by LDC firms? Probably by shedding their labour intensive parts (by outsourcing or leaving the business) and product upgrading (differentiation). Following an example makes the argument less abstract. The television receiver industry is taken as an example. It is expositionally easy to consider an industry where matters have settled.[49]

It is first of all necessary to state that the television-receiver industry in the 1960s consisted of a few firms that made complete television sets, from plug to screen so to speak. Around 1970 Japan began penetrating the US market for colour sets. When the Japanese found their products excluded from the market by several trade-policy measures, other Asian countries took over. How did American firms react to increased foreign competition? First of all they advocated protection, but besides that, they followed 'a competitive strategy centred on reducing product prices through the transfer of labour-intensive processes to low wage countries' (OECD, 1985, p. 163).

What does this essentially mean? And what does it mean for statistical tractability? Employment in the television receiver industry declined by almost 30 per cent between 1977 and 1981. The average skill-intensity in the industry increased, as skilled-intensive parts of the production process were relocated to low-wage countries. We make an additional assumption, probably one that is not too hard to defend: we assume that the plants relocated to low-wage countries were the relatively less R&D intensive ones.

This brief history of the television-receiver industry indicates that 4-digit level data do not guarantee that production facilities can be treated as an entity. This means that what the statistics reveal as a change in production technique might in reality be a relocation of certain parts of production capacity under pressure of foreign competition. Or stated again differently (this is an important point) even at a 4-digit level industry characteristics might change over time and might differ from the foreign counterparts, without being caused by technological change.

Accepting the assumption of non-homogeneous goods has serious implications. Rejecting a trade-related explanation of observed trends on the account of the 'wrong' changes in factor intensity in industries, as Krugman and Lawrence (1993), is probably less convincing after this story (see Box 2.2).[50] See Nahuis (1999) for an alternative graphical comparison of the hypotheses advocated here and the Stolper-Samuelson mechanism. The available direct evidence on complementarities between skill and technology or capital fits in this story also. The *non*-R&D (unskilled labour) intensive parts of production are relocated to low-wage countries, so skill-upgrading in the developed countries and R&D activity go hand in hand.

The story above indicates that this type of outsourcing of unskilled-labour intensive parts of industries (due to non-homogeneity of production) even seems relevant at a high level of disaggregation. In that case it seems that a babel like confusion between advocates of the asymmetrical technological development and the trade-based explanation is not to be excluded.

To summarize the argument: we argue that production within a single industry cannot be viewed as one homogeneous activity. Therefore outsourcing can be relevant. One can similarly argue that the level of disaggregation is insufficient to label the industry with a single production function, hence that a single indicator for skill intensity is misleading.

The magnitude of outsourcing is often stated to be too small (see for example Lawrence (1994)). Usually, however, the only outsourced activities of multinational firms taken into account are those that are reimported again *internally* as intermediate inputs. The definition of outsourcing embraced here is more broad. Here we propose to take 'involuntary outsourcing' also into account and to include imported goods sold under the brand name of a domestic firm.[51] This extended definition implies that closed plants, formerly producing 'parts' of the spectrum of goods produced *within* a single industry, are part of outsourcing.

BOX 2.2 DISAGGREGATION, NON-HOMOGENEOUS GOODS AND OUTSOURCING

This box illustrates the outsourcing hypothesis in the Lerner-Pearce diagram. Now, for expositional ease, the following fixed-proportion technology is assumed:

$$F_S = \min\left(\frac{L_S}{b_{SL}}, \frac{H_S}{b_{SH}}\right), \qquad F_U = \min\left(\frac{L_U}{b_{UL}}, \frac{H_U}{b_{UH}}\right)$$
$$\text{with } \frac{b_{SH}}{b_{SL}} > \frac{b_{UH}}{b_{UL}}, \tag{2.8}$$

where b denotes the unit factor requirement. Figure B2.2a shows the initial equilibrium (H/L is the endowment point; vectors from the origin denote the allocation of factors over the two sectors).

Outsourcing
A change (in technology, transport costs or tariffs) makes it possible to slice up the value chain in high-skill high-value added parts and low-skill low-value added parts. Assume that assembly of the different parts is costless and that the LDC is incapable of producing high-value added parts. The division of productivity in high-skilled (denote by h) and low-skilled intensive parts (denoted by l) implies:

$$\frac{b_{SH}^h}{b_{SL}^h} > \frac{b_{SH}}{b_{SL}} > \frac{b_{SH}^l}{b_{SL}^l}, \qquad \frac{b_{UH}^h}{b_{UL}^h} > \frac{b_{UH}}{b_{UL}} > \frac{b_{UH}^l}{b_{UL}^l}. \tag{2.9}$$

The value added assumptions imply that $P_S^h > P_S > P_S^l$, $P_U^h > P_U > P_U^l$. The graphical representation is given in Figure B2.2b. The low-value added goods can no longer be produced profitably in the developed country (following the same procedure as in panel (a) one can infer that in panel (b) goods markets still clear). Hence, the new w_L/w_H is lower than previously and both goods are produced with a more skilled-intensive technique.

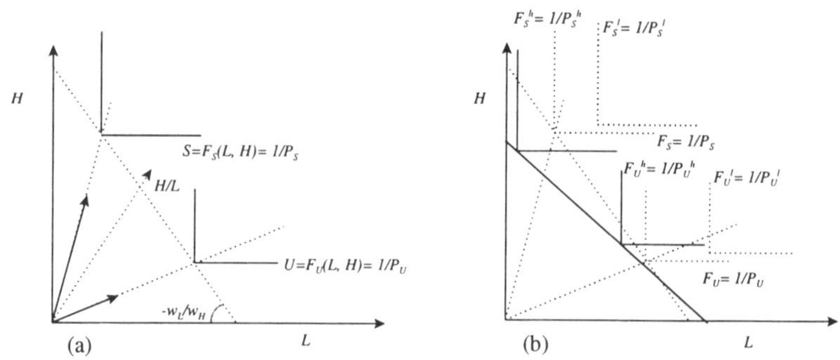

Figure B2.2 *Outsourcing*

Direct evidence

Supporting evidence for this story can be found in numerous case studies (see note 49). Cooper (1994) analyses the very unskilled-intensive (by wage profile) industries, notably textiles, apparel and leather (TAL). Rising prices of the TAL goods relative to all manufactured goods are found. Can this be explained in the face of stiff competition from LDCs? Only when one allows for product upgrading or outsourcing this can be explained. The price of imported goods might decline while the price of the slightly different domestic goods rises (see Figure B2.2).

Though indicative, we are well aware that this piece of evidence is not discriminating between the trade induced changes that we are emphasizing here and biased technological change.

Disaggregation, non-homogeneous goods and outsourcing: theory and evidence

Wood (1994) focuses on the north and the south, not the US alone. His comprehensive work starts with an exposition of his theoretical framework. In the Heckscher-Ohlin framework the essential (that is, in the long run immobile) production factors are unskilled labour and skill and knowledge attached to people. Technology is seen as prior used skill. In contrast with most other authors, Wood distinguishes three types of labour: those with no education, those with basic education, and a heterogenous group of skilled workers. Workers without any education are assumed to be located in the south only, and irrelevant for manufacturing. So the standard Heckscher-Ohlin model is applied with two factors: workers who have basic education and those who have more skills. Basically, Wood applies the FCT method. Earlier FCT calculations used the wrong method: they underestimate effects because inappropriate factor-input coefficients are used. The coefficients should be adapted for the fact that they have changed as a consequence of the increase in competition. The most labour intensive sub-industries have long disappeared. Part of trade is therefore now 'non-competing', in the sense that these goods are not produced in the north, as they are unprofitable by their high unskilled-worker intensity. Therefore, usual calculations are strongly downward biased. This consideration is similar to the argument set forth above.

In the empirical part Wood takes this potential bias into account. A countervailing bias in the estimation is the result of the fact that prices of the 'non-competing' goods, now produced in the south, will probably be lower. This causes an increase in demand for the good. Wood's own calculations correct for both biases and result in an estimate of a cumulative negative effect on demand for basic-educated workers of about 5 per cent of the total number of basic-educated workers in the economy.

Next Wood adjusts his calculations for the following facts: trade in services has been 'forgotten' and the effect of less skilled-intensive goods production on the skill composition of intermediate non-traded goods should also be taken into account. Doing this yields a doubling of the estimate. Another doubling of the estimate is required to account for defensive labour-saving innovation. To be clear about the latter doubling of the estimate, trade/competition induced labour-saving technology change is meant. Wood states: 'Their search has been spurred by the most powerful of all instincts – the urge to survive' (p. 159).[52] In sum, a relative demand shift of 22 per cent results for those with more skills to those with basic education. Trade matters!

There are some serious limitations to the work of Wood. It is obvious that multiplying the original estimate by four is rather arbitrary. Besides, the adjustments to the FCT approach are not very transparent nor objective and they do not meet the critics on using FCT expressed earlier.

A point raised by Baldwin (1995) is that Wood assumes similar technology, in a TFP sense, and therefore overstates considerably the number of workers with basic education required for producing the now imported goods in the north. Lawrence (1994) is also critical. He tries to ridicule the adjustments of the coefficients proposed by Wood and states (p. 14): 'Suppose *all* the growth in US imports over the 1980s reflects imports of products that where *not* produced in the United States in 1980 at all. Had imports from developing countries not increased, therefore, Americans would have spent their money on *other* domestic products... which were not unusually labour-intensive.' Lawrence did not recognize the point that 'non-competing' imports are not a static fact of the economy. To illustrate this, we reiterate Wood's statement: 'Northern firms have *ceased* to produce many manufactured items currently imported from the South.' (p. 73, italics added).

Feenstra and Hanson (1995) presented a stylized augmented Heckscher-Ohlin model that rigorously represents the argument set forth above. They, however, change more assumptions at once: besides dropping the 'homogeneity-of-goods' assumption they allow for mobile capital.

Feenstra and Hanson (1995) notice the incompatibility of the relative wage trends in the US and Mexico and the observed changes in factor intensities in production in the US (see also Section 4.6). Instead of concluding that asymmetric technological change is the principal cause they develop an augmented Heckscher-Ohlin model.[53] The model is the Dornbusch-Fischer-Samuelson (1977) model, amended to allow for unskilled and skilled workers and capital. Final products are produced from a continuum of differentiated intermediate goods that are tradable. These intermediate goods can be ordered by increasing skill intensity. The division of production of intermediate inputs over the countries in the relatively high and the relatively low skilled is determined by four 'endowments': the level of technological attainment (TFP), the capital endowment and of course the skilled and unskilled workers. The essential assumption is that imported goods from an industry are (completely) different from domestically produced ones. Hence assuming an equal production technology across countries within a sector is thought to be too restrictive.[54]

In the Feenstra and Hanson model foreign direct investment (FDI) or an improvement in the technological abilities in a low-skilled-abundant country will increase the relative wage of skilled workers in *both* countries. This result is easily understood intuitively. The less-skilled-intensive part of the whole spectrum of intermediate goods shifts from the skilled rich to the skilled-poor country; relative demand for skilled workers increases in the skilled-rich country. For the skilled-poor country these new intermediaries are relatively skilled-intensive ones, thus also in this country the relative demand for skilled workers increases.

Part of the empirical contribution of the paper concentrates on the magnitude of foreign direct investment. Usually approaches based on FDI or outsourcing are nipped in the bud by stressing that the share of imports in purchased materials is too small, or that the flows of FDI are insignificant in magnitude.[55] Feenstra and Hanson argue that the employed definitions for outsourcing are too narrow to assess the actual impact of the possibility not to produce at home. The usual definition includes only materials, parts, components and so on. One example might clarify why this definition, excluding finished products and contract work, is too narrow. General Electric now imports all its microwaves (from the Korean firm Samsung) and sells these under their own brand name; this practice is excluded by the usual definition for outsourcing. So Feenstra and Hanson propose to amend the definition and to include all intermediate and final goods used in production or sold under a domestic brand name. Foreign direct investment in Mexico increased between 1983 and 1987 by almost 1000 per cent. Using the proposed definition, they are able to explain 25 per cent to 33 per cent of the wage-bill share change of non-production workers by changes in the import shares.[56] In Feenstra and Hanson (1996) new data are used and up to 50 per cent of the wage-bill share increase of non-production workers can be attributed to outsourcing.

A limitation to Feenstra and Hanson's work is that the reported regression is not an actual test for their model but only a fairly general test of the correlation between imported 'inputs' and the wage-bill share of high-skilled workers.

Lawrence (1994) states: 'If outsourcing is important, the decline in blue-collar intensity in the US should be associated with an increase in blue-collar intensity abroad' (p. 23); the outsourcing model he has in mind is not clear, but clearly different from the Feenstra and Hanson model. And he continues (p. 23): 'if global changes in technology were dominant, we should see parallel increases in the ratio of blue to white collar employment in the US and in the rest of the world and similar movements in wages'. Taking the Feenstra-Hansen model into account this is not an obvious repercussion of a global change in technology. Here we would suggest again that it might be the case that a kind of babel like confusion is going on between those in favour of the biased-technological-change explanation (that is hard to separate from trade-induced outsourcing) and those favouring a significant role for trade.

6.2 Imperfect competition and rents: A new assumption

Perfect competition is a common assumption in the models relating trade to wages. A notable exception to this extreme assumption is Borjas and Ramey (1993). They argue that competition in parts of the economy is far from perfect and that the consequent profits are shared with workers.[57]

Imperfect competition and rents: theory and evidence

Borjas and Ramey (1993) analyse what variables have the same long run trend as the wage inequality series from 1968 to 1988. The only variable that does a good job is the trade deficit in durable goods. R&D expenditures and de-unionization had very different trends in earlier periods. The intuition is that durable goods industries are thought to be highly concentrated and therefore to generate high rents which are shared with their (in majority less-skilled) workers. They provide a theoretical underpinning of this result. In a Cournotian two-sector model, rents are dissipated in case of increased competition (proxied by an increase in the number of competing firms). The concentrated industries are relatively important employers of less-skilled workers. Therefore, a relatively small increase in foreign competition is likely to have a large effect on wage inequality.

The theory is tested by applying the model to different regions in the US. The magnitude of the inequality increase differs substantially between regions (a sufficient immobility of workers is implied). In a regression[58] the strong correlation between manufacturing employment and skill return vanishes after controlling for time and area. After disaggregating manufacturing according to concentration and vulnerability to foreign competition, a strong negative correlation between high-concentration import industries and the return to skill is found.[59] Borjas and Ramey also pose the obvious question of whether the magnitude of the estimated coefficients is large enough to explain the observed trend. Ten per cent of the wage inequality change can be explained by the change in the share of employment in high-concentration industries.

6.3 More factors: Capital as a relevant factor of production

Usually, the standard trade theory is presented in terms of capital and labour. Today capital does not seem to be that important any more. Even *mentioning* capital in the theoretical explanation is not necessary for Krugman and Lawrence (1993). Wood (1994) argues that capital is reproducible and tradable and states that 'all theorists would agree that this type of input cannot in itself affect the pattern of trade' (p. 34). In itself the argument might be correct, but we are looking at 'a kind of opening-up'. Sachs and Shatz (1994) and Feenstra and Hanson (1995) disagree with the denial of the role for capital and suggest adding an initial lack of physical capital in the LDC to a similar framework as presented in the discussion of the Stolper-

Samuelson theorem. Now two sectors are distinguished, one producing with capital and skilled workers and one with capital and unskilled workers. The LDCs initially lack capital and skilled workers. Opening-up would yield a capital flow from the US to the low-wage country resulting in the US only in a relative decline in the reward of the unskilled workers and increased imports of the unskilled-intensive good (thus, only the weak version of prediction (1) and prediction (4) in Section 4). This is not too hard to see. Facilitating possibilities for capital flows result in a flow from the US to the LDC. That no skill intensity change occurs is obvious as the construction of the model does not allow for it. Also a relative goods price change is not necessary. These issues are important, because the rejection of Heckscher-Ohlin theory (and, often, with it the relevance of trade) is based on one of the predictions (2), (3) and (5) not coming true.

Unfortunately empirical work with Sachs and Shatz's model in mind is not available.

6.4 Many goods, many sectors

Specialization

The Stolper-Samuelson predictions in the case of specialization are not unambiguous. Though it is beyond the scope of this chapter to discuss the method used by Leamer (1987) to determine the Stolper-Samuelson derivatives, it might be useful to stress the basic message.[60] The basic message bearing on the Stolper-Samuelson theorem in a 3 x n (factor x goods) model is that the results hinge on the specialization pattern of the country under consideration. Or, stated differently, the logic of the Stolper-Samuelson theorem ceases to be rectilinear if: '(a) some non-produced goods rise in price relative to produced goods and some fall, and/or (b) the prices of produced goods change relative to each other and relative to some non-produced goods' prices' (Ethier, 1984, p. 145). This need not be a theoretical exception recalling the remarks in Section 6.1. There we argued that certain goods within narrowly defined industries ceased to be produced. This would imply that complete specialization within industries on slightly differentiated goods or on stages of the production is not unlikely. Empirical work taking this modification explicitly into account is not available.

More sectors: a sheltered sector

Advanced economies usually incorporate a sector sheltered from foreign competition that amounts to at least half the economy. Does taking the sheltered sector into account explicitly alter the predictions?

In theory,[61] adding a non-traded goods sector does not alter the basic results. As long as there are at least as many goods as factors produced in both countries, factor prices are derived from zero-profit conditions and the price of the non-traded good adjusts correspondingly. In case of specialization the strict logic is again broken. Leamer (1996a) also discusses two cases. In the first case (without specialization) international prices set factor prices via zero-profit conditions, the marginal demand for labour is 'external'. In the second case the economy specializes in one traded good and the non-traded good and the marginal demand for labour is said to be 'internal' (and thus a trade deficit, for example, matters).

This last case is probably implicit in Cooper's[62] mind when arguing: 'labour is shed from the tradable sector even with flexible wages (because goods are non- homogeneous, specialisation is very likely).[63] The question then becomes: How good is the economic system at absorbing the released labour with little decline in wages?' This is a highly relevant question if the manufacturing size declines substantially; how much of a wage decline do workers have to accept to be absorbed in the non-traded sector.

7. REMAINING ISSUES

Our main focus is on the empirical assessment of potential explanations for the increased education premium. There are, however, some approaches that deserve attention although the ideas offered are not empirically scrutinized. The remainder of this section provides an impression of possible interesting ways to proceed (in empirical or theoretical work).

Bhagwati and Dehejia (1994) reject the Heckscher-Ohlin theory as the assumptions are judged to be unrealistic. They provide three alternative theories that are not yet empirically examined. The impact of trade on the domestic economy is different from that in Heckscher-Ohlin theory, where the level of the terms of trade is crucial. The three arguments, by which a more integrated world causes inequality, all start from the basic premise that the increased integration of the world causes a higher volatility in the terms of trade. The first mechanism runs from higher volatility (more sudden shifts in comparative advantage on a sectoral level) to higher labour turnover to more frictional unemployment and a flatter earnings growth curve due to less skill accumulation by learning by doing. Bhagwati et al. call this the 'rolling stones gather no moss' approach. If this works out asymmetrically between high- and low-skilled workers, as higher skills are less firm specific, an increased wage differential results. A second approach takes a similar starting-point (increased uncertainty) but focuses on education, that is the supply of skilled versus unskilled workers. Becoming skilled requires sunk investment costs. Increased uncertainty makes workers less reluctant to upskill and enlarges the sustainable wage differential. We have already seen, however, that supply changes cannot tell the whole story. Finally and more conventionally, in a dual economy with only one unionized sector 'good' jobs are lost by terms of trade changes that negatively affect the competitiveness of the unionized sector. In the US the impact of competitiveness on union power is reflected in a smaller intersectoral wage differential and in Europe in a decline of the number of insiders. Bound and Johnson (1992) have already shown that this mechanism is not very relevant.

In the previous section we argued that the assumption of identical or homogeneous production functions within specific industries over time and across borders is too strong. Rosen (1981) explains increasing inequality by arguing that the assumption of homogeneous production factors within specific subcategories on the labour market is too strong.

The 'superstar' model, illustrating his argument, builds on two basic blocks, increasing returns to quality and decreasing production costs to scale. Some results: the larger the market the more important returns to scale are

(Rosen connects these with increased communication and transport possibilities) and this produces more skewed income distributions. Rosen provides therefore an explanation for the increase in overall inequality. When there is some asymmetry for different skill types in this mechanism, this provides a potential explanation for the increase in the relative wage of high-skilled workers. To illustrate the asymmetry we have in mind: think of a surgeon and someone doing assemblage. Assume a type A and a type B of the surgeon and the assembly worker. Surgeon[64] A gives you 10 per cent additional chance to survive compared to B and the assemblage worker A is 10 per cent quicker than B. You are probably willing to pay 10 per cent more for assemblage worker A than B, but probably 50 per cent more for surgeon A than for B.

Reich's book on *The Work of Nations* (1992) discusses growing inequality of workers with different types of jobs. Empirical work is not discussed explicitly, but as Reich seems to have strong ideas about the topic it is worth discussing part of the book.

Convincingly, Reich stresses that the traditional job classification utilized by the Census Bureau is not suited for analysing issues connected with ongoing globalization. It is argued that the employer type is no longer relevant for analysing future prosperity but that the type of work that is performed is what matters. A new categorization is proposed: workers carrying out routine production services, in-person services and symbolic-analytic services. The routine production workers are workers performing routine jobs, like blue-collar workers and low- and middle-level managers and data processors. The in-person service workers perform also routine jobs, but the nature of the service requires personal interaction. Symbolic-analysts perform complex jobs in solving and identifying problems. Think of researchers, certain lawyers and accountants and so on. The latter two categories of workers are rising in number while the former is declining. Reich foresees a quick globalization: 'In a *very few* years, there will be virtually no way to distinguish one national economy from another' (p. 172, italics added). This process, directed by declining transport and communication costs, is evaluated as to have inherent advantages for symbolic-analysts, as their jobs are most highly rewarded in the world. The routine production workers see their jobs relocated to Asia, Latin America and so on. The in-person service workers are protected largely from global competition but they face increased competition from former routine production workers and (il)legal immigrants. Besides the indirect competition labour-saving technology also is negatively affecting the prosperity of in-person service workers. But in this part of the economy new jobs are generated easily.

Reich tells a story (rhetorically excellent) of FPE: transport and communication costs are negligible and symbolic-analysts are relatively scarce in the world. Essentially he endorses the trade explanation.

8. GENERAL DISCUSSION AND CONCLUSION

It is time to evaluate the evidence presented. The facts are indisputable. Despite the increasing relative supply of skilled workers the evolution of factor rewards has been beneficial for the more skilled. Thus, somehow an explanation needs to be given for the fact that factor demand has developed in a direction favourable to skilled workers.

It might be necessary to repeat the aim of this chapter. There is a whole palette of potential explanations for the increase in the relative wage of the high skilled: education, trade, immigration, de-unionization, technological change, deregulation and so on. We want to sort out how much room there is for both the technology-based explanation as well as the trade-based explanation. It appears that trade has been taken of the table as a potential explanation. We argue no more than that a significant role for trade cannot be excluded on the basis of the available evidence. Thus we put trade as a potential explanation back on the palette of explanations alongside others; including unfavourable technological change for low-skilled workers. To corroborate this general conclusion, several arguments will be put forward. First to be stressed is why we are not fully convinced by technology as the dominant explanation in itself. Second, subsequent arguments for rejecting the role for trade are disputed.

The first reason for not being eager to accept the 'technology related' explanation is the following. Technology is just what we observe as an unexplained residual. Even one of the opponents to the trade explanation is critical with respect to the offered alternative; Krugman (1994b, p. 148): 'If trade doesn't explain the rise in inequality, we are left primarily with the all-purpose explanation known as "technology".' There is no need to push this any further. In the remainder of the discussion will be set forth like it is between the *opponents of the trade explanation* and *opponents of the technological change explanation*.

The second reason for being hesitant in accepting technological change as the relevant alternative is the low discriminative power of the offered direct tests, like the ones by Bartel and Lichtenberg (1987), Mincer (1991) and Berman et al. (1994). In Section 6.1 we argued that the significant relation between skill-upgrading and technology might equally well reflect outsourcing of low skill/low tech elements of the production process.

A third issue is the unlikeliness of the biased technological change explanation viewed in the light of the sluggish productivity increases since the early 1970s. Is there a reason for weak productivity growth to be biased strongly against unskilled workers?[65] Summarizing the arguments: technological change is a residual 'explanation' and what is claimed to be evidence for technological change might equally well favour the trade explanation, as outsourcing and biased technological change are to a considerable extent observationally equivalent.

Next we will reconsider subsequent arguments that were brought on the scene by opponents of the trade hypothesis.

It is argued that in the 1980s the shift in factor demand was for a vast part a shift that took place within single industries, indicating that trade-induced goods demand changes had not come into play (Katz and Murphy, 1992; Berman, Bound and Griliches, 1994). For two reasons this argument is not convincing. First, and most important, as claimed above (see Section 6.1), the common statistical classification of the data is not suited to distinguishing between and within industry shifts. Second, as stressed by Baldwin (1995) and Baldwin and Cain (1997), the issue at hand should be analysed in a general equilibrium framework. That proves that within- and between-industry shifts occur both in the case of technological change-induced and trade-induced reallocation.

Furthermore there are tests related to the magnitude of international trade relations. The volume of trade, an argument used by Krugman and Lawrence (1993), is rather uninformative. As stressed by Bhagwati (1991) ex post trade volumes are not informative in assessing the role for trade. If the situation on the world market changed dramatically it can be that despite (and due to) huge internal adjustments, trade volumes have not changed. Related calculations, using the FCT approach, are not very informative from the perspective of the Heckscher-Ohlin model. The Stolper-Samuelson theorem connects factor prices with international price changes. Quantity changes do not enter the formulation of the theorem. The FCT is determined by the equilibrium conditions for the factor markets (Leamer, 1995). Furthermore, deviations from balanced trade are also muddling up FCT calculations.

Next, there are tests that are actually clearly related to the Stolper-Samuelson theorem: for example the tests on factor intensities within industries, shifts in employment between industries and goods prices. We can summarize our arguments on the unreliability of these tests of the Stolper-Samuelson theorem as follows. First, the habit of treating industries – that are classified for statistical purposes by goods type – as a homogeneous entity with single factor-input requirements is obfuscating the analysis. Second, as the Stolper-Samuelson theorem is formulated with unchanged endowments it is necessary to keep in mind that this has not been the case. And finally we want to emphasize the difficulties in measurement once again (for example, measuring skill). Next it is briefly explained how the summarized arguments here might muddle up the detection of the Stolper-Samuelson predictions in the data.

Summarizing: the two research directions followed after the disappointing results with the basic 2 x 2 x 2 Heckscher-Ohlin model – searching for evidence for biased technological change and amending the trade model – lead to the following conclusions. First, there is still room for both the trade and the biased-technological-change explanation. The evidence in favour of the trade explanation is flimsy. We pointed out, however, that several conceptual difficulties complicate proper measurement. The most important difficulty is in distinguishing trade-induced outsourcing from biased technological change. A possible solution to this deadlock is to analyse the role for trade in a general equilibrium model; an approach we will follow in Chapter 8. The difficulty in distinguishing the outsourcing explanation from the biased-technological-change explanation also hampers the trust one can have in the evidence favouring the latter explanation. Moreover, most literature pushing the biased-technological-change explanation does so by default. This type of explanation – actually an 'explanation' based on ignorance – would gain considerably in power if the bias in technology could be explained itself. This is an issue to which we turn in the Chapters 3, 4 and 6. In these chapters we will develop structural models that explain how a bias in technological change comes about.

NOTES

1. Numbers in the introduction are taken from Lawrence and Slaughter (1993).
2. This issue will only be touched upon in Section 7.
3. In summarizing, the focus is on the results; the method and data will be mentioned only in passing or in notes. For technical details refer to the original papers.
4. The focus is on white full-time employed men's average hourly wage rate in 1964–90 from the Current Populations Survey.
5. The strong rise in unemployment in the beginning of the 1980s is thought to be a potential explanation.
6. Full-time male and female wage and salary workers who worked at least one week are in the sample for 1963–87. The method: supply and demand framework where each distinct category is treated as a distinct labour input with a separated (though via substitutability interacting) labour market.
7. An unequal substitution elasticity is required between the non-labour production factor and the distinguished types of labour.
8. Assigning changes in demand caused by trade as between-industry shifts is not correct from a general equilibrium perspective. See Section 4 on trade theory.

9. To give some indication, Borjas and Ramey (1993) use time series of a cross-section of cities. Their finding is that a 10 percentage point increase in the foreign born population lowers the relative wage of high school dropouts to college graduates with more than 5 per cent. They use Current Population Survey data, so this does probably not tell much about illegal immigrants. They just report that immigrants increase the supply of less skilled workers.
10. In the US it is a common notion that high-school quality deteriorated. The notion is corroborated by the data (Bishop, 1991). Levy and Murnane (1992) argue that this cannot be an important part of the explanation, as the education premium also rose for older workers, who acquired their education much earlier. A second reason why the hypothesis is not convincing is that the hypothesis also requires that education (above high school) compensates for the deficiencies in secondary school, as the college graduates also passed through high school.
11. The basic 2 x 2 x 2 HO model is built on the following assumptions: identical constant returns to scale production functions, free trade, no specialization (factor endowments are not that dissimilar in that they remain inside the cone of diversification), perfect competition on all markets, factors are fixed in supply and preferences are identical and homothetic. A lot of these assumptions are not met in the real world. This is hardly surprising, but makes the theory not useless as a 'tool to think'. Bhagwati and Dehejia (1994) spend many pages on arguing the inadequacy of this model.
12. The logic of the Stolper-Samuelson theorem remains valid in case of the monopolistic competition (see Helpman and Krugman, 1985). Ethier (1984) treats higher dimensions than 2 x 2 and concludes: 'Nevertheless it is fair to conclude that the Stolper-Samuelson theorem is not fundamentally altered by an increase in the number of goods.' (p. 145). In the case of specialization the Stolper-Samuelson theorem does not hold in general. Komiya (1967) analyses HOS theory in case a non-traded goods sector is added. In that case it is still trustworthy to think in Stolper-Samuelson terms.
13. They restrict themselves to the manufacturing sector in the period 1979–89.
14. In fact the logic of the Stolper-Samuelson theorem also holds with fixed factor input ratios. Leamer (1994) stresses this but immediately admits that *if* factor substitution is feasible low-skilled workers are substituted for high-skilled workers. The empirically relevant case is a substitution elasticity significantly different from zero. Therefore, this theoretical argument receives no further attention.
15. In his comment on Lawrence and Slaughter (1993), Hall raised the point that the continuing supply increase might muddle up the picture. A solution to this problem is provided: consider the development of skill intensity within sectors compared to the national average.
16. The Rybczynski theorem shows that factor intensities are unaffected by factor-endowment changes. The theorem is, however, only relevant for small countries.
17. The difficulty in identifying skilled and unskilled workers is problematic no matter how the classification is approximated. This, however, has to be taken for granted and will not be underlined over and over again in the remainder.

18. Katz and Murphy (1992) assume that trade affects production workers disproportionally. This issue will be discussed extensively in Section 6. Correcting for this yields that trade is significantly affecting the unfavourable development of the demand for unskilled workers.
19. They use the Annual Survey of Manufactures from 1978 to 1990. They distinguish 131 industries and 150 trade partners. Again, skill intensity is represented by production versus non-production workers.
20. Deardorff and Hakura (1994) state that this indecisiveness of Murphy and Welch confuses them; Murphy and Welch show that trade is a significant determinant of changes in wages so the unexplained part must be something else.
21. See also Bhagwati and Dehejia (1994).
22. The issue, also raised by Bhagwati and Dehejia (1994), that it is highly problematic to calculate the FCT in case of serious trade imbalances is not disputed; this makes it troublesome to use the FCT approach for the 1980s in any case.
23. See Section 4.5.
24. Deardorff and Hakura (1994) is an excellent reference to formulate the right questions.
25. Ignoring Krugman's (1995b) critique, reported on in the previous section.
26. Usual practice is to account for the globalization effect and attribute everything else to technology. Leamer takes the mirror image.
27. Why is the actual timing earlier? The used non-production versus production workers division is incorrect. Criticizing this division implies that Leamer has to come up with an alternative. High- and low-wage workers are constructed by marking the highest paying industry as the high-wage (that is, high-skill) industry. Next it is assumed that only high-wage workers are employed in that sector and the same for the lowest-paying industry. The ratio of high- and low-wage workers in all other sectors is constructed by linear intrapolation.
28. Several sources are used; see Table 1 in the paper.
29. To overcome the idea that Mexico might be very special due to location and its taking part in the NAFTA we refer to Currie and Harrison (1994) who find a very similar pattern in Morocco after abolishing several trade restrictions, cited via Harrison and Revenga (1995).
30. One change that might be messing up the 'laboratory experiment' is the relaxation of minimum- wage policy (since 1983). This might be relevant for relative wages. Bell (1994), however, analysed micro-data to measure compliance of minimum wages. The minimum wage though was a constraint for only 2 per cent of the plants. The minimum-wage policy is therefore further ignored. Hanson and Harrison (1995) document the considerable lowering of import licence requirements and tariffs besides the freeing of export possibilities (and a devaluation of the nominal exchange rate)
31. Mexico's exports (in value) of industry products (in 1985) was directed towards the US for over 90 per cent (in 1992 almost 80 per cent). For imports the number for 1985 is 76 per cent and for 1992 is 65 per cent (source, Wildcat data system, CPB).

32. Like assuming that Mexico is relatively skilled labour abundant to the rest of the world (but trade with the rest of the world is only a minor part of Mexican trade).
33. The question on the cause of the initial change in relative wages is of course highly relevant. If we do not answer this, the second line of reasoning is circular.
34. The consistency of these stories should of course be checked in formalizing the arguments.
35. Davis and Topel (1993) provide a nice analogy with respect to the need for more direct evidence. Assuming that biased technological change is causing the increased inequality is like inferring the existence of Pluto because Neptune's orbit does otherwise not fit in the theory. The existence of Pluto is much more compelling ever since it was actually sighted through a telescope. Thus an apparatus which can actually trace the bias in technology is needed for the argument to be convincing.
36. To control for the possibility that the computer dummy accidentally catches the fact that workers might be able to capture rents, separate regressions are run for union and non-union workers. Surprising (and yet unexplained) is the result that the return to computer use for unionized workers is only about one third of that for non-unionized workers.
37. There is serious reason for suspicion as it is the use of e-mail that appears most productive.
38. The data on white males from one Current Population Survey in each year from 1963 to 1987 are used.
39. International competition, proxied by the ratio of net export to GNP, is however significantly negative; the right direction for a trade induced explanation.
40. Their research relies on data from the Annual Survey of Manufacturers from 1959 to 1989. The data are from producers, therefore all inputs used in production are available. Data are disaggregated at the 4-digit level.
41. For example, the between-industry shifts in employment are decomposed in shifts due to more or less net exports, more or less employment occupied for defence build-up and the employment active to produce for domestic consumption. Employment is approximated by output data.
42. They decompose this in plant size and equipment; the actual complementarity is between equipment and skill.
43. The data from Lawrence and Slaughter, that Leamer criticized, are used in some (preliminary) regressions, estimating the impact of technological change (affecting low skilled workers only) on wages, leaving prices constant.
44. A regression is run with the relative goods price change as the dependent variable and factor shares are independent variables. Industries are the units of observation. The estimated coefficients are relative factor price changes (this relation is easily derived from zero profit conditions (mobility of factors across industries and CRS technology is assumed)). Combining these estimates with actual observations on endowments, output, relative and absolute wages and the skill intensity of production, they can exclude certain explanations as being dominant.

45. Baldwin and Cain, however, do not make a clear choice between a boundary at 11 or 12 years of education. They arbitrarily shift this boundary to provide the most convincing underpinning of their explanations.
46. Population Survey data for 1973, 1979 and 1988 are divided into 32 sub-samples by education (4), experience (4) and sex (2). For each sub-period they estimate an extended wage equation and calculate from the coefficients the real wage for all (32) cells. In proceeding like this they filter out well known differentials caused by race, industry and region.
47. They assume full employment despite that they have chosen to leave certain types of workers out of the sample. They assume constant and equal elasticity of factor substitution between all workers and all industries. So, it as easy to substitute for a college graduated manager with 30 years of experience, a similar employee with 25 years of experience, as a high school drop out with no experience. Elaborating further that this is a strong assumption is probably not necessary.
48. The idea is not new, and can be found in discussions on the relevance of models for intra-industry trade. See, for example, Finger (1975).
49. This example is taken from the *Costs and Benefits of Protection* study by the OECD (1985), refer to this for more examples.
50. For the story to fit the facts, one should be aware of the declining manufacturing sector employment. It is necessary, or at least plausible, that the trends in this story are accompanied by an on average shrinking manufacturing sector. Besides that, the increased skill intensity in manufacturing has to be met with a response somewhere else in the economy, as it is not solely supply driven. Somewhere the domestic unskilled should be absorbed. Either the composition of the group of unemployed should become more unskilled, or the sheltered sector should show an opposite skill intensity development. The minimum requirement for this story not to be rejected is that the skill intensity of the sheltered sector is increasing less than the one in the exposed part of the economy, as the relative supply increase of skilled workers has to be taken into account.
51. Feenstra and Hanson (1995, 1996) propose this latter extension.
52. Sachs and Shatz test for the assertion that low-skilled workers are shed from low-skilled intensive sectors (this would show up erroneously in TFP), and indeed find a higher TFP growth in low-skilled intensive industries. This would 'add another few percentage points to the overall negative effects of trade on employment' (p. 33). But they are unsure whether this is caused by increased import competition.
53. Markusen and Venables (1995) build a model with similar features, but a less transparent and more ad hoc specified one. The model is driven by endogenous location decisions and the possibility of outsourcing by multinationals. Their model is also able, if FDI is allowed for, to generate similar results for relative wages as the Feenstra Hanson model.
54. To get some feeling for this deviating modelling approach it is useful to reiterate some crucial equations: production of the final good is modelled as costless assembly of foreign and domestic intermediates, according to a Cobb-Douglas production function:

$$\ln Y = \int_0^1 \alpha(z) \ln x(z)\, dz \quad with \quad \int_0^1 \alpha(z)\, dz = 1$$

and intermediate inputs are produced (domestically or abroad) by:

$$x(z) = A_i \left[\min \left(\frac{L(z)}{a_L(z)}, \frac{H(z)}{a_H(z)} \right) \right]^{\psi} [K(z)]^{1 - \psi}$$

where z ε [0,1].

55. See Lawrence (1994).
56. The average skill intensity of plants in Mexico remained almost constant and the variance of the skill intensity increased. This is exactly what the model predicts, see Feenstra and Hanson (1995). Feenstra and Hansen (1995) also derive a 'modified Stolper-Samuelson result'; for trade to play a role in increasing the wage differential the domestic price should rise more than the import price. Empirically this is clearly confirmed.
57. Bound and Johnson (1992) also consider industry-rents shared with workers. They, however, apply a closed economy framework and judge the evaporation of rents as caused by de-unionization (that is, a trade-independent explanation).
58. A Mincerian equation is estimated using CPS data for 1976 to 1990 divided over 44 metropolitan areas. From this, standardized skill differentials are obtained. These are regressed on industrial composition (the manufacturing employment share), a vector of area and period fixed effects, the rates of unemployment and foreign born workers in the population and female participation rates.
59. These highly concentrated (import) industries, like primary metals, construction equipment, office equipment, aircraft and parts and so on, were in 1976 the important employers of less-educated workers. This percentage showed a steep decline alongside their wages towards 1990.
60. Chapter 8 shows the procedure to derive Stolper-Samuelson derivatives with three production factors.
61. See for example Komyia (1967).
62. In a comment on Krugman (1995a).
63. The remark in parentheses is also from Cooper but added in this quote.
64. Think of lawyers and accountants if you wish.
65. Krugman (1995b) tests whether the technological explanation is internally consistent given the total factor productivity growth over the last 20 years. The conclusion is that the technology explanation is feasible. Chapter 4 corroborates this and shows that it is actually likely that a bias in technological change is accompanied by sluggish productivity growth.

PART II

Theory

3. A GPT in a research and assimilation model: Exploring wage dynamics (I)

1. INTRODUCTION

Computers are now used in the production process of virtually every good or service. Moreover, production of numerous new goods and services has been made possible by the advances in computer technology. The pervasiveness of computers makes one wonder whether the invention of the semi-conductor, the heart of computer technology, did mark the beginning of a new industrial revolution. However, despite the omnipresence of information technology (IT),[1] the 'revolution' does not seem to have brought a revolution in productivity development. Or, to summarize the so-called productivity paradox: 'You can see the computer age everywhere but in the productivity statistics' (Solow, 1987).[2]

Did the IT revolution lead to increased inequality? Though the massive introduction of computers did not bring spectacular productivity gains, it seems to have lead to increased inequality on the labour market. The increase in inequality between skilled and unskilled workers is an undisputed empirical phenomenon. To explain this phenomenon the three main suspects are education, trade with low-wage countries and technological change. Educational attainment has increased, so that explanation seems to go in the wrong direction.[3] Trade does not seem to account for a large part of increased inequality.[4] So technology is left to explain the remainder. The obvious candidates causing the bias in technology are computers and IT.[5] So computers seem to have increased inequality without paying off in terms of productivity. Is there a relation between these observations?

The story this chapter tells is simple. The computer is not simply a new gadget that is installed and improves productivity. The computer technology opens new opportunities with respect to the organization of work and the innovative process. The full benefits of the IT revolution are not realized immediately. It takes time and resources to see and learn about the possibilities the new technology offers. Up to now, the computer still imitates a paper-oriented culture and discoveries of new opportunities are still being made. Implementation of such a technology, which is called a General Purpose Technology (GPT), takes place throughout the economy and typically generates a cycle. For analysing such a learning process it is useful to think about the process driving technological progress in a more subtle way than the one-dimensional perspective that usually suffices.[6] Imagine a firm where a part of the workforce has tasks with explicit learning possibilities and hence these workers can generate improvements in production technology.[7] What happens when a technology like the semi-conductor/computer arrives? First, the R&D workers assimilate the opportunities the technology offers, that is, for example, office clerks play with e-mail and surf the Internet to find out what is in it for them. Once the R&D workers assimilate the opportunities the new technology offers, this investment pays off in more efficient development of firm-specific applications of the GPT. For example, e-mail turns out to be an efficient way of communicating and the Internet appears to be a productivity enhancing source of information for office clerks. So what goes on is that a GPT initially distracts R&D workers from directly productive R&D, and causes growth to slow down. But, once the opportunities of the GPT are recognized, research is more effective and growth accelerates. Moreover, assuming that skilled workers have an advantage in research, learning and assimilation implies that the introduction of computers increases the relative wage of skilled workers too (Bartel and Lichtenberg, 1987). Hence the two observations are related.

The purpose of this chapter is to argue that with a more realistic formulation of the learning process the introduction of a GPT generates a slowdown in productivity growth without (large) fluctuations in R&D labour (as these are empirically not observed). Moreover, the chapter demonstrates that sluggish productivity growth and increased wage inequality might be related phenomena. Actually the chapter shows that the two observations are the consequence of one and the same thing: a GPT.

The chapter proceeds as follows. The next section discusses the literature on GPTs and growth cycles. In Section 3 the learning process outlined above is embedded in a model with skilled and unskilled workers. Section 4 analyses the steady-state properties of the model. Section 5 analyses the long-run impact of a GPT, whereas in Section 6 the impact effects of a GPT are analysed numerically with a calibrated version of the model. Section 7 concludes.

2. GENERAL PURPOSE TECHNOLOGIES

Economic growth's most important driving force is technological progress. Technological progress in turn seems to be driven by a few major technical or organizational breakthroughs. Examples are, the concept of a factory as a way of organizing work, the steam engine, electricity and the transistor. These concepts, mostly breakthroughs in engineering, have turned out to be widely applicable throughout the economy. Application of such a generic function in a specific context requires investment. Most of the applications of what later turned out to be general concepts have never come to the minds of the original inventors. Only slowly the possibilities were recognized and the GPTs themselves improved over time. Some common features of GPTs can be distilled from the rich descriptive material provided by David (1989, 1990 and 1991). David compares the general purpose engine of the previous *fin de siecle*, the electrical engine, with the current one, information technology, and finds remarkable parallels. Analogous are the sluggish labour productivity growth at the turn of the century, the constancy of real wages and the fact that a new technology was introduced everywhere but did not contribute to recorded productivity growth. As the macro-economic tendencies at the end of the previous century seem similar to today's, we might learn from understanding why no productivity gains from early electrification were recorded.[8] First, it took some 20 to 30 years before the adoption of electrical engines was substantial. Second, early adopters used electricity driven systems that were backed up by mechanical power derived from steam or water.[9] The third explanation is crucial for the aspects of technological change we focus on. It turned out that substantive productivity gains of electrification were only accomplished once it was recognized that factories could be designed in a previously unthinkable way and hence work could be organized much more efficiently. Illustrative evidence for this claim is that early applications of electrical engines were used to lift water back up to the top of the water-wheel. While at the same time keeping the factory organized by the restriction that all machines needed to be connected by belts to the single power source. David (1989, p. 23) argues that 'The advantages of the unit drive [that is for every machine a separate power source, RN] for factory design were manifold, extending well beyond the savings in inputs of fuel...'. Hence, an appropriate way of looking at a new GPT might be that it fuels innovation.[10]

The key ingredients of our model, that copes with the diffusion process of GPTs, as described by David (1991) and Bresnahan and Trajtenberg (1995), are the following. The first characteristic of a GPT is that it is potentially beneficial to all firms. In the model a GPT generates new possibilities for innovation in all firms. The second characteristic of a GPT is that firms have to invest in assimilating the nature of the new GPT before they can put effort in developing useful applications of the GPT. Hence we model assimilation of new ideas out of a pool of knowledge (including a GPT and spillovers of other firms' research activity) that yields a stock of accumulation capabilities. This stock is a necessary input to research for directly applicable knowledge. For this type of research to remain a viable activity, the stock of accumulation capabilities should grow. For an earlier application of the two-stage research structure, see Rustichini and Schmitz (1991).[11] Finally, fruitful application of the GPT by a specific firm generates spillovers that may interact with the GPT. This indirectly implies that an inherent potential for improvements of the GPT exists (that is the third characteristic David attributes to GPTs). We examine cases with and without a positive interaction effect. Finally, the model distinguishes two types of workers: skilled workers that are fully specialized in research and unskilled workers who only produce.

2.1 Related literature

The dynamics of a GPT have recently been analysed formally by Helpman and Trajtenberg (1994, further HT). They develop a general equilibrium model where GPTs require complementary inputs before they can be applied profitably in the production process. Complementary inputs developed for a previous GPT are incompatible with a newly arrived GPT. The invention of complementary inputs requires a fixed labour input. The arrival of subsequent GPTs causes cycles. A typical cycle consists of two phases, a phase where firms produce final goods with the old GPT and components are being developed for the new GPT, and a second phase where final goods producers switch to the new GPT and the development of components for that GPT is continued. A consequence of such a technology cycle is that GDP declines in the first phase, as workers switch from production to research to invent new inputs, and increases again in the second phase once the new technology is implemented.[12, 13]

Our analysis is in the spirit of the work by HT but deviates in the mechanisms driving the application of a new GPT and differs importantly in the empirical implications. The mechanism this chapter introduces is new in the literature on GPTs and growth cycles. The first difference is that in our model existing firms have to cope with the new GPT instead of new firms as in HT. In this sense our analysis is complementary to HT's.

The second difference is that we analyse the process of implementation instead of the decision to adopt. Several approaches using vintage-type models to analyse the decision to adopt have strong similarities with each other. For example, adoption of a technology in vintage-capital models generates growth cycles due to the assumption that the starting level of expertise in a new vintage depends negatively on the pace of technological progress. The IT revolution is seen as a positive productivity growth shock in investment-specific technology, see Greenwood and Yorukoglu (1997) and Yorukoglu (1998).[14] Helpman and Rangel (1999) provide a different perspective by analysing the decision of workers to adopt a new technology. Workers are heterogeneous with respect to their experience with a previously dominant technology. Similar to the vintage-capital models, a slump occurs if the efficiency of workers who start using the new technology is lower than with the previous technology. Further, a slump is more likely when a larger proportion of the workforce switches (the latter is positively related to the learning speed with the new technology). Slightly different again, heterogeneity in firms' productivity, combined with learning by doing in the capital-goods producing sector (instead of workers using the technology), could possibly generate a slump (Felli and Ortalo-Magné, 1997). A slump is not necessary, however, as the most productive firms switch immediately and increase output. This counteracts the fact that the least productive firms lower investment in the old technology. If the first effect outweighs the second, the arrival of a new technology will be followed by a boom.[15]

Third, the R&D process in our model differs from the literature. In our model research increases productivity as is common in the literature. Moreover, assimilation activity to comprehend new ideas is required to keep up the research potential. On the one hand this two-stage learning process resembles other approaches with two different types of R&D, as R&D-labour can perform different types of research. However, our approach contrasts these approaches in the sense that in other approaches firms have to decide either to aim at breakthroughs or to aim at improvements, as in our model the research activities are complements instead of substitutes (Cheng and Dinopoulos, 1996).[16]

Fourth, this chapter integrates the income-inequality puzzle with the productivity-paradox puzzle. Greenwood and Yorukoglu (1997), exceptionally, also analyse these phenomena in a unified framework but they have a very different perspective on the arrival of a GPT, as stressed above.[17]

Finally, this chapter adds to the literature by analysing the short and long-run impact of different *types* of GPTs; GPTs that interact with the existing knowledge stock and GPTs that do not.

The (empirical) implications of our model are substantially different from the existing literature. First, no reallocation from R&D to production is required to generate a slump in productivity growth. The empirical implication of the HT approach is that the occurrence of growth cycles should be accompanied by strong fluctuations in resources devoted to R&D. However: 'fluctuation in research and development and in employment of resources are not large [...] enough to explain significant and progressive fluctuations in output' (Felli and Ortalo-Magné, 1997, p. 4).[18] Andolfatto and MacDonald (1998) also need too large fluctuations in R&D labour to replicate post-WWII data in a model where technologies that differ in quality and difficulty to acquire are innovated or assimilated. Our model shows that it is possible to generate growth cycles in an R&D model without large fluctuations, actually no fluctuations at all, in R&D labour. Secondly, our structure generates a slowdown in output growth, along the introduction of a superior technology, without the necessity to assume creative destruction, forgetting, or some kind of incompatibilities.

3. THE MODEL

The impact of the emergence of a new general purpose technology will be analysed in a general equilibrium model.

3.1 Description of the model

The model economy consists of a 'large' number of firms, N, that each produce a differentiated variety of a consumption good. Firms are located equally spaced on a circle. The closer they are the more similar their knowledge. The economy is populated with H skilled workers and L unskilled workers who both supply their labour fully inelastic. We will consider the case of symmetric industries. In a symmetric allocation every firm can allocate *L/N* and *H/N* workers; denote the former L_N and the latter H_N. The remainder shows that the analysis might be expressed in terms of the representative firm and consumer as the number of firms does not play a role. Consider a representative firm *i*, indexed $i \in \{1,...,N\}$. Firm *i* produces good *i* with a linear production technology,[19]

$$x_i = f_i L_i \ . \tag{3.1}$$

Unskilled labour, *L*, produces good *x* with productivity *f*. Notice, for later reference, that the production technology features constant returns in the traditional production factor, labour. The stock of productive knowledge is firm-specific in the sense that only firm *i* is able to produce with knowledge stock f_i.[20] Firm *i* accumulates knowledge by investing in R&D using the following technology:

$$\dot{f}_i = A_f (f_i H_{Ri})^{1-\alpha} k_i^{\alpha} \ , \ 0 < \alpha < 1 \ . \tag{3.2}$$

A_f is a research productivity parameter, H_R is skilled labour devoted to research activity and *k* is the stock of accumulation capabilities. Several features of this specification are worth noticing. First, the technology for accumulating productive knowledge exhibits decreasing returns to *f*. This implies that relying fully on internally generated experience (*f*) is insufficient to keep growth going.[21] Hence, as the amount of labour is assumed to be fixed, to have growth that does not peter out, accumulation capabilities should grow. The second feature is that there are decreasing returns to labour in knowledge production. That is, unlike the production technology for goods, a replication argument does not hold here. Hence, given the amount of assets (*f* and *k*), doubling the amount of labour at a point in time does not lead to a twofold increase in the flow of new ideas or productive knowledge (cf. Jones, 1995b).[22]

Accumulation capabilities serve as an asset in the research process for new productive knowledge. These capabilities are accumulated according to:

$$\dot{k}_i = A_k \left[\left(\frac{Z}{k_i} \right)^{\varphi} k_i \right]^{\lambda} H_{Ai} \;, \; \varphi > 0 \,, \lambda \leq 1 \;, \tag{3.3}$$

where A_k is the parameter governing the productivity of the assimilation technology and H_A is skilled labour engaged in assimilation. Z is the knowledge pool available to a firm, this knowledge pool is 'filled' with the latest generation of the GPT and knowledge spillovers. Spillovers are the more general principles, developed as a byproduct of research for firm-specific knowledge, that are indirectly useful for other firms. As only symmetric equilibria are considered, we can, without loss of generality, assume that a single general knowledge pool exists, hence no firm index is added. The term between parenthesis, Z/k, is the 'learning potential' or 'knowledge gap': the size of the knowledge pool relative to the firm's accumulation capability. As φ is positive, a larger learning potential implies more effective assimilation. Consider two cases with respect to φ. First, if $\varphi > 1$, 'fishing out' applies. The most effective ideas, or equivalently the most obvious ideas, are 'fished out' of the common knowledge pool first. In that case the more accumulation capabilities a firm has, the harder or less effective further assimilation will be ($\partial \dot{k} / \partial k < 0$). An alternative hypothesis is that firms learn to assimilate, reflected in $\varphi < 1$. Then more accumulation capabilities imply easier assimilation; $\partial \dot{k} / \partial k > 0$. The specification features decreasing returns in Z and k together if λ is less than unity. One might expect that assimilating existing knowledge is an activity where doubling the amount of human input leads to doubling the output.[23] Therefore, the specification for assimilation again reflects the replication argument. Figure 3.1 provides an overview of the activities of firms 1 to N. For firm 1 the dashed arrow indicates the knowledge spillover generated by research that implements the GPT.

In the knowledge pool, spillovers from productive knowledge of other firms and the GPT play a role.[24] The knowledge pool with the GPT of generation j looks like:

$$Z(j)=Z(f_1,\ldots f_N,Q(j)) \quad . \tag{3.4}$$

Note that $Z(j)$ changes gradually over time due to changes in f_i for all i. $Q(j)$ is the effective quality of the GPT of generation j. We assume $\partial Z/\partial f_i$ and

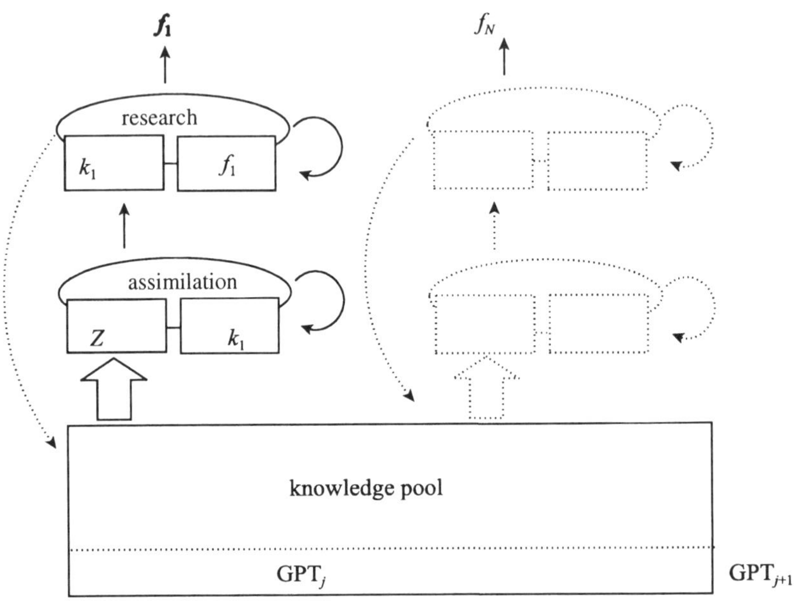

Figure 3.1 Structure of the model

$\partial Z/\partial Q$ >0. The invention of a next generation GPT occurs serendipiditously and hence is unexpected and does not require resources. A more extensive discussion of the function Z is postponed until Section 5.

Table 3.1 Producer behaviour

Producer *i*:

$$\max\ V_i = \int_0^\infty e^{-rt}\left(x_{it}p_{it} - (H_{Rit}+H_{Ait})w_{Ht} - L_{it}w_{Lt}\right)dt \tag{3.5}$$

subject to: (3.1), (3.2), (3.3), (3.15) and $H_{Ri} \geq 0$, $H_{Ai} \geq 0$.

First-order conditions (f.o.c.) are:

$$\frac{\partial \mathcal{H}}{\partial L_i} = 0 \qquad p_{xi} = \frac{\varepsilon}{\varepsilon - 1}\frac{w_L}{f_i} \tag{3.6}$$

$$\frac{\partial \mathcal{H}}{\partial H_{Ri}} = 0 \qquad (1-\alpha)\, q_{fi} A_f H_{Ri}^{-\alpha} f_i^{1-\alpha} k_i^{\alpha} = w_H \tag{3.7}$$

$$\frac{\partial \mathcal{H}}{\partial H_{Ai}} = 0 \qquad q_{ki} A_k \left[Z^{\varphi} k_i^{1-\varphi}\right]^{\lambda} = w_H \tag{3.8}$$

$$\frac{\partial \mathcal{H}}{\partial f_i} + \dot{q}_{fi} = r q_{f_i} \qquad \frac{w_L L_i}{f_i} + (1-\alpha)\, q_{fi} A_f f_i^{-\alpha} H_{Ri}^{1-\alpha} k_i^{\alpha} + \dot{q}_{fi} = r q_{fi} \tag{3.9}$$

$$\frac{\partial \mathcal{H}}{\partial k_i} + \dot{q}_{ki} = r q_i \qquad \alpha q_{fi} A_f (f_i H_{Ri})^{1-\alpha} k_i^{\alpha-1} + \lambda(1-\varphi)\, q_{ki} A_k Z^{\varphi\lambda} k_i^{\lambda-1-\varphi} H_{Ai} + \dot{q}_{ki} = r q_{ki} \tag{3.10}$$

Producer behaviour is summarized in Table 3.1. Producers maximize the value function V indicating the present value of the firm, subject to the technical constraints discussed above and a downward sloping demand curve familiar to preferences with goods that are imperfect substitutes. The Hamiltonian of the formulated maximization program is denoted by $\mathcal{H}$. We focus on interior solutions, hence the inequality constraints are not binding (for a discussion of corner solutions, see Appendix C).

First-order condition (3.6), derived from the optimal use of unskilled labour or the optimal supply of output, shows that firms set a mark-up over the unit labour cost. The mark-up is inversely related to the price elasticity, ε. Optimal allocation of skilled workers in both the research and the assimilation activity requires the marginal cost, w_H, to equal marginal return (the l.h.s. of equations (3.7) and (3.8)). The marginal product of skilled workers is the marginal increase in the productive knowledge stock, valued at the shadow price, q_f. The l.h.s. of (3.8) shows the marginal product of skilled workers in assimilation; the shadow price q_k times the marginal addition of accumulation capabilities. The marginal return of assimilation is increasing in Z and increasing (decreasing) in k if φ smaller (larger) than unity. The no-arbitrage condition (3.9) says that the return of investing q_f in the financial market should equal the return of investing in productive knowledge. The latter consists of three parts. The first part on the l.h.s. of equation (3.9) is the direct benefit of a marginal increase in productive knowledge: the marginal increase in the value of production, $\partial xp_x/\partial f$.[25] The second term is the increase in the knowledge base, again valued at the shadow price. The third term is the capital gain term. The second no-arbitrage condition, (3.10), has a completely analogous structure. The first term on the l.h.s. is the direct benefit of a marginal increase in accumulation capabilities, that is, the increase in the value of productive knowledge, $\partial \dot{f}q_f/\partial k$. The second term is the change in 'fishing potency'. If 'fishing out' applies (φ>1) this term is negative, indicating that accumulation of k today implies *ceteris paribus* a lower return to assimilation tomorrow. The third term is again a capital gain term.

Preferences and consumer behaviour are standard and presented in Table 3.2. Maximizing the CRRA utility function subject to the wealth accumulation constraint yields the familiar Ramsey rule. θ and $1/\rho$ denote subsequently the pure rate of time preference and the elasticity of intertemporal substitution. In the second stage of the budgeting problem, consumers decide on the division of their spending over different varieties at each point in time. A downward-sloping demand schedule for each variety (equation (3.17)) results. Symmetry of firms, hence prices, results in a uniform distribution of spending over the differentiated goods.

Consumer preferences, presented in Table 3.2, imply that variety as such does not play a role.[26]

In order to stress that no reallocation of workers from production to research is needed to generate growth cycles, we analyse the special case without any substitution possibilities between (skilled) research workers and (unskilled) production workers. Hence, the labour market is completely segmented. Aside from expositional ease, the segmentation can be motivated by the notion that skilled workers have a comparative advantage in research and assimilation and unskilled workers in production. The qualitative results, we believe, would not be affected by relaxing this assumption. The segmentation of the labour market implies the following equilibrium condition:

$$H_N = H_R + H_A \ , \tag{3.11}$$

for skilled workers. And similarly for unskilled workers:

$$L_N = L \ . \tag{3.12}$$

Solution of the model and characterization of the steady state will be the topic of the next section.

Table 3.2 Consumer behaviour[a, b]

$$\max\ U=\int_0^\infty e^{-\theta_t}\frac{1}{1-\rho}X_t^{1-\rho}dt \tag{3.13}$$

$$\text{where } X_t=N\left(\frac{1}{N}\sum_{i=1}^{N}x_{it}^{\frac{\varepsilon-1}{\varepsilon}}\right)^{\frac{\varepsilon}{\varepsilon-1}} \tag{3.14}$$

Yields:

$$\frac{\dot X}{X}=\frac{1}{\rho}\left((r-\dot P_X/P_X)-\theta\right) \tag{3.15}$$

Where:

$$P_X=\left(\frac{1}{N}\sum_{i=1}^{N}p_{xi}^{1-\varepsilon}\right)^{\frac{1}{1-\varepsilon}} \tag{3.16}$$

$$p_{xi}=P_X\left(\frac{x_i}{X}\right)^{-\frac{1}{\varepsilon}} \tag{3.17}$$

Notes: [a] For a discussion of the financial market and equilibrium asset holdings in this type of models see van de Klundert and Smulders (1997).
[b] In symmetric equilibrium: $P_X=p_x$. and $XP_X=Nxp_x$; we normalise N to 1, hence $X=x$.

4. THE STEADY STATE

Because this section deals with the steady state, we abstain from discussing the introduction of a new GPT. Section 4.1 defines and Section 4.2 solves for the steady state. The determination of relative wages and labour market equilibrium is discussed in Section 4.3.

4.1 Definition of the steady state

A steady-state equilibrium is defined as a path where all variables grow at a constant, possibly different, rate and where the allocation of labour is time-invariant. It is easy to show that Z should be homogeneous of degree $(\varphi\lambda+1-\lambda)/\varphi\lambda$ in average productive knowledge to have positive steady-state growth (use (3.2) and (3.3)). If $\lambda<1$ this implies that the knowledge pool should grow at a higher rate than the knowledge generated by agents in the economy. Empirical research should answer the question whether such a relation is plausible. For the remainder of the analysis the specification is specialized to one with constant returns in assimilation and the knowledge pool Z, by setting $\lambda=1$. So, Z should have a long-run growth rate equal to that of f:

$$Z(j)=fZ\left(1,\frac{Q(j)}{f}\right) . \tag{3.18}$$

Keep in mind that there is no steady growth in the GPT, hence in the long run the arguments of Z tend to a constant which is denoted z in the remainder. Finally we normalize prices to one. The balanced growth path is characterized as (using (3.6), (3.7), (3.8) and (3.9)):

$$\frac{\dot{X}}{X}=\frac{\dot{x}}{x}=\frac{\dot{f}}{f}=\frac{\dot{k}}{k}=\frac{\dot{Z}}{Z}\equiv g . \tag{3.19}$$

Thus all variables grow at a common constant rate, denoted g. For prices holds that:

$$p_x=1 \Rightarrow \frac{\dot{P}_X}{P_X}=\frac{\dot{p}_x}{p_x}=0,\ \frac{\dot{q}_f}{q_f}=\frac{\dot{q}_k}{q_k}=0,\ \frac{\dot{w}_L}{w_L}=\frac{\dot{w}_H}{w_H}=g . \tag{3.20}$$

4.2 Solution of the model

Some additional notation simplifies the exposition further. Define $u=H_R/H_N$ as the fraction of skilled labour doing research. Labour market equilibrium for high skilled workers can now be written as (using (3.2) and (3.3))[27]:

$$R=\left(\frac{A_f}{A_k}z^{-\varphi}H_N^{-\alpha}\frac{u^{1-\alpha}}{1-u}\right)^{\frac{1}{\varphi+\alpha}}. \qquad \text{(LAB)}$$

Where $R\equiv f/k$, that is the steady-state ratio of productive knowledge and accumulation capabilities. Multiplying both sides of the definition for R with z, that is constant in the steady state, shows that R is proportional to the knowledge gap (Z/k).

Having defined the steady state by (3.19) and (3.20), the Ramsey rule (3.13) can be written as $g=(1/\rho)(r-\theta)$. To find the steady-state growth rate, the equilibrium rate of return on savings should be substituted in this expression. Firms equate the return to investment in productive knowledge and to investment in accumulation capabilities. Hence, the rate of return can be found using only one no-arbitrage condition. Combining the no-arbitrage condition for accumulation capabilities (3.10) with the two static optimality conditions ((3.7),(3.8)) yields a return of:

$$r=(1-\varphi)A_k(zR)^{\varphi}(1-u)H_N+\frac{\alpha}{1-\alpha}\frac{u}{1-u}A_f(uH_N)^{1-\alpha}(R)^{-\alpha}. \qquad (3.21)$$

The return is a weighted sum of accumulation equations (3.2) and (3.3) and should be read as the demand for capital. To derive an explicit expression for g turns out to be cumbersome, therefore we rely on a graphical approach. Rewriting (3.2) to (use the definition of R): $g=A_f(uH_N)^{1-\alpha}R^{-\alpha}$, and substituting this into the Ramsey-rule yields an expression for the required return on savings by consumers. Combining this with (3.21) yields an equation for the capital market equilibrium (CAP):

$$R=\left(\frac{H_N A_k z^{\varphi}}{\theta}\left[\left(\frac{\alpha}{1-\alpha}-(1-\rho-\varphi)\right)u+(1-\rho-\varphi)\right]\right)^{-\frac{1}{\varphi}}. \qquad \text{(CAP)}$$

To show that for certain parameters an equilibrium with positive growth exists, a graph in the u-R space is convenient, see Figure 3.2.[28] The LAB-curve is upward sloping; if u is high, that is most skilled workers are allocated to accumulate productive knowledge, the knowledge gap is large. The CAP-curve is downward sloping. The intuition is that if the knowledge gap is large, R high, the real return in assimilation is high (hence u is small to close the knowledge gap). The equilibrium allocation and the ratio of productive knowledge to accumulation capabilities is found at the intersection of the CAP and LAB curve. The growth rate is computed by equation (3.2) (use the definition of R and u).

4.3 Relative wage determination

The model has a recursive structure in the sense that the solution for the growth rate does not require solving the complete model. This is easily seen as u and R are determined by the intersection of the CAP and LAB curve that are both independent of the number of unskilled workers (L_N). Thus, the endowment of unskilled workers, and the relative wage do not affect the growth rate.[29] This section provides a digression on relative wage determination.

Equating the no-arbitrage conditions (3.9) and (3.10), taking into account the definition of the steady state one can derive an expression for the relative wage ($\omega \equiv w_L/w_H$):

$$\omega=\frac{1}{1-\alpha}\frac{H_N}{L_N}u\left(\frac{\alpha}{(1-\alpha)}\frac{u}{(1-u)}+\alpha-\varphi\right), \qquad (3.22)$$

where (3.7) and (3.8) are substituted for the shadow prices. As L_N does not affect u, ω is decreasing in the endowment of unskilled workers.

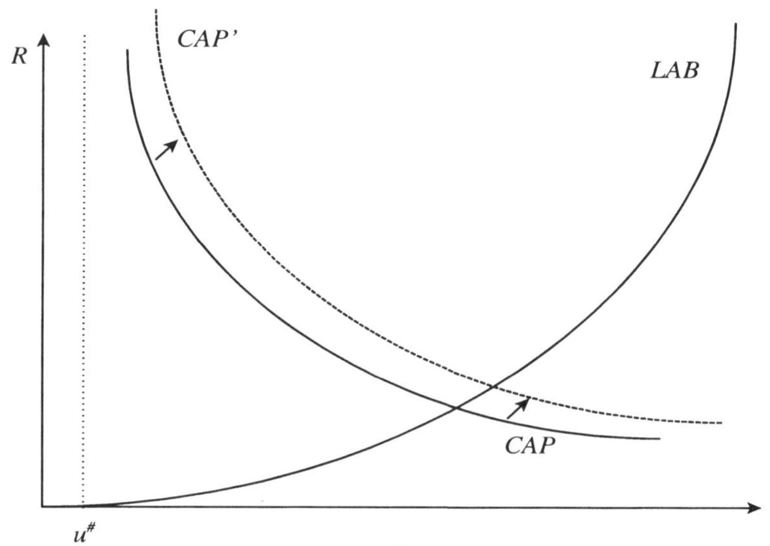

Figure 3.2 General equilibrium

To show the working of the wage determination in the model, consider the following comparative static experiment. Assume consumers become less patient, so the rate of time preference (θ) increases. The CAP-curve shifts up (CAP' in Figure 3.2) and the LAB-curve is not affected. The new steady state is characterized by a larger knowledge gap and a higher fraction of the workforce doing research (u). From (3.22) it is easily seen that the relative wage of low skilled workers increases in u. What is the intuition for the positive relation between the relative wage of unskilled workers and the degree of impatience? Unskilled workers are specialized in final goods production. At a point in time, production possibilities are fixed as both the stock of productive knowledge and the supply of unskilled labour is fixed. Dynamically, however, the increase in relative wages is consistent with less patience. An increase in the relative wage increases the return to investment in research (see (3.9)) and does not affect the return to assimilation. Therefore the return to the asset that has a direct impact on productivity is increased relative to the return to the asset that affects only future accumulation. Therefore, a higher fraction of the workforce is doing research in the new steady state.

The impact of a higher rate of time preference on growth however yields an ambiguous result. Graphical inspection shows that more skilled labour is allocated towards research but with a lower productivity, as the ratio of productive knowledge to accumulation capabilities (*R*) is lower. By linearizing the model around the steady state we can derive that an increase in the rate of time preference turns out to decrease growth unambiguously (for details and other comparative statics, see Appendix A and B).[30]

5. A NEW GENERAL PURPOSE TECHNOLOGY

This section has two purposes. First, the relation between the arrival of a new GPT and productive knowledge will be discussed. Second, the impact of an increase in the GPT's quality (*Q*) and the long-run performance of the economy are examined.

So far the knowledge pool is defined as:

$$Z(j)=Z(f_1,...f_N,Q(j)) \quad . \tag{3.4}$$

The previous section established that in the long run the knowledge pool should grow with the growth rate of f to have a steady state with positive growth. At time t, a GPT of quality j arrives; the arrival of a new GPT is exogenous. We analyse a one time unanticipated arrival of a GPT. We assume that at the arrival of GPT j the economy is on a stationary path, as discussed above. Furthermore, the GPT's quality is one dimensional and a new GPT is strictly better than the previous one.

The interaction of a GPT with knowledge spillovers from other firms that are available in the knowledge pool gives rise to two cases. In the first, a GPT is simply new knowledge that is widely applicable and requires time and effort to implement. In the second case, a GPT is generating ideas and a positive interaction between spillovers and a GPT exists. In the remainder the intuition for and the long-run impact of the two cases will be discussed.[31]

In case (i) no interaction exists between the new GPT and spillovers from productive knowledge, hence $Z_{fQ}=0$ (where the subscript indicates the cross derivative). An example of a functional form where the cross derivative is zero is the case where f and Q enter additively in Z. So:

$$Z(j)=\bar{f}+Q(j) \quad , \tag{3.4A}$$

where $\bar{f}$ is the effective spillover of firms' productive knowledge. The equivalent of this equation in words is: the new GPT is pervasive, as it affects learning possibilities in all N industries and it requires complementary investments to advance the performance of the technology in a specific environment. However, the technology does not positively interact with technologies available now and in the future. More concrete such technologies are gadgets that turn up everywhere but whose technology does not make other technologies more productive, examples are most office supplies and improved materials. What about the long run impact of such a technology? The definition of a steady state (see Section 4.1) requires the ratio of productive knowledge (f) and accumulation capabilities (k) to be constant in the steady state. Both f and k grow in the steady state, so the impact of the new GPT ceases in the long run.[32]

In case (ii), the ideas generating GPT, there is a positive interaction of the new GPT with existing and future productive knowledge, hence $Z_{fQ}>0$. The GPT of our times, semi-conductor technology, seems to fit well in this classification. Semi-conductors do seem to improve the efficiency of all devices already developed. To infer the impact of such a GPT on long run growth we specialize Z to:

$$Z(j)=\bar{f}Q(j) \quad . \tag{3.4M}$$

In this case fruitful application of the GPT by firms increases their productive knowledge which in turn generates spillovers that interact positively with the GPT. This mechanism implies that successful applications 'improve' the GPT. This is the third characteristic of a GPT David (1989) described (see Section 2). The impact on long run growth is inferred from the linearized version of the model. The impact of an increase in the quality of the GPT is:

$$\tilde{g} = \Upsilon \varphi \tilde{Q} \ . \tag{3.23}$$

Variables with a tilde denote deviations from a steady state, hence $\tilde{x} \equiv dx/x$. Υ is always positive. Hence, an increase in the GPT's quality leads to higher long-run growth.[33]

6. DYNAMICS AND CALIBRATION

The previous sections showed conditions for a new GPT to increase the long-run rate of growth. However, the short-run impact of a new GPT might be different. This section analyses the initial impact of a new GPT and the transitional dynamics. To that end we derive the dynamic equations of the non-linear model, calibrate the model and analyse the transitional dynamics.

6.1 Dynamic equations

In Sections 4.2 and 4.3 reduced form equations for the model are derived in terms of the relative wage (ω), research allocation and the knowledge gap. Differential equations for these variables describe the dynamic behaviour of the economy, one equation for the state-like variable R, that is the ratio of two state variables (f and k), and two equations for the jump variables u and ω.

$$\dot{R} = R\left(A_f (uH_N)^{1-\alpha} R^{-\alpha} - A_k (zR)^{\varphi} (1-u) H_N\right) , \tag{3.24}$$

$$\dot{u} = \frac{u}{\alpha}\left(\frac{\alpha}{1-\alpha}u + \alpha(1-u)\right)H_N A_k (zR)^{\varphi} - \frac{u}{\alpha}\left((1-\alpha)\omega\frac{L_N}{H_N}\frac{1}{u} + \varphi\right)A_f (uH_N)^{1-\alpha}R^{-\alpha}\,, \tag{3.25}$$

$$\dot{\omega} = \omega\left((1-\varphi-\rho)A_f(uH_N)^{1-\alpha}R^{-\alpha} - \theta + \frac{\alpha}{1-\alpha}(uH_N)A_k(zR)^{\varphi}\right)\,. \tag{3.26}$$

The dynamic equation for the knowledge gap is easily found by log differentiating the definition of R and substituting the accumulation equations for productive knowledge and accumulation capabilities. The second dynamic equation is found by log differentiating the conditions for the optimal static allocation of skilled workers, (3.8)=(3.7), and solving for the evolution of the relative shadow price q_f / q_k by the two no-arbitrage conditions. Log differentiating (3.6), (3.7) and (3.8) and combining these equations in levels with (3.9) yields, after some manipulation, the last dynamic equation.

6.2 Calibration

The real-life data we want to explain are the growth rate of GDP and the relative wage of unskilled (high school or less education) versus skilled (college educated) workers in the 1980s for the US. GDP growth for the US is close to 3 per cent annually. In the simulation we use 3 per cent. The relative wage is approximately 0.73 in 1979 (see Davis, 1992).

The following parameters are used in the simulations. OECD (1993) reports that 17 per cent of the population is college educated, so in our definition 83 per cent is unskilled in 1980. Hence we use a ratio for L_N/H_N of 5. A common value for the rate of time preference is 0.05. The inverse of the intertemporal rate of substitution ρ is taken to equal 1, to limit the number of cases to be considered. For *z*, the level of the GPT, we use 0.1 and 1. The productivity of assimilation is set to one. 1-α indicates the importance of the past experience in own productive knowledge accumulation and indicates the degree of decreasing returns to skilled labour in knowledge production. For 1-α estimations are available varying from 0.17 to 0.38.[34] For α we take 0.5, 0.6 and 0.7. Finally φ and the productivity of research (A_f) remain to fit the model to generate the desired growth rate and relative wage.

Table 3.3 Calibration

		α					
		0.5		0.6		0.7	
z	0.1	φ = 3.76	A_f = 0.082	*φ = 1.05*	*A_f = 0.035*	*φ = 0.34*	*A_f = 5e-4*
	1	φ = 3.76	A_f = 0.026	*φ = 1.05*	*A_f = 0.009*	φ = 0.34	A_f = 0.001

Notes: The other parameter values are: H_N=1, L_N=5, A_k=1, ρ=1 and θ=0.05.

Table 3.3 shows that depending on the size of the decreasing returns in research, a fishing-out result emerges (φ>1) if α is relatively small. With higher αs a lower level of φ is able to replicate the data.

The steady states which are in italic in Table 3.3 will be used as starting point for the dynamic analysis in the next section (hence we distinguish one case with a φ larger than unity and one with a φ less than unity).

6.3 Dynamic analysis of a GPT with an application to the wage-inequality debate

The purpose of this section is twofold. First, we want to show that the model is able to generate a slowdown in the growth rate following the arrival of a new GPT. Second, we confront the predictions of the model with the data on the wage-inequality debate. Numerical analyses are carried out with a shooting routine. As the model has one state variable and two jump variables the model is globally saddle-path stable with one negative and two positive roots. All numerical examples considered fulfil this requirement.[35]

Case (i) A new GPT

In Case (i) Q and f enter additively into the knowledge pool, see equation (3.4A). In this case, the economy returns to the original steady state (see Section 5). Figure 3.3 panel (b) shows the growth rate after the arrival of a new GPT.[36] The arrival of a new GPT enlarges the knowledge pool (the dashed line in panel (a)) and makes it relatively attractive to invest in assimilation.[37] So, less is invested in research to accumulate directly productive knowledge and at the arrival of a new GPT the growth rate jumps down.[38] The assimilation effort lowers R and hence increases the effectiveness of research. So, after the above-steady-state effort in assimilation, research intensity is increased and hence growth rises and the assimilation of the GPT pays off. In the long run, the GPT has a negligible

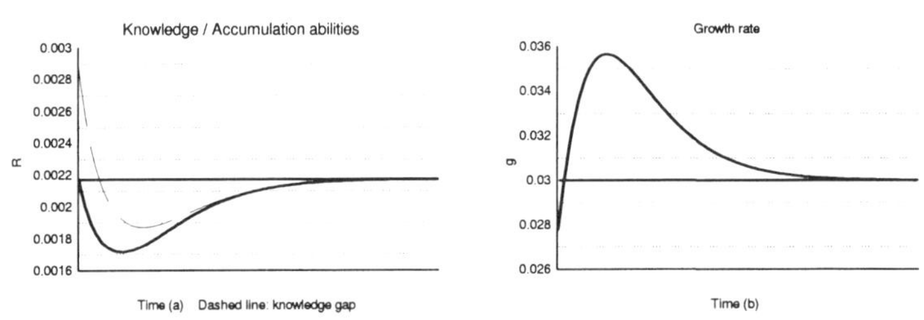

Figure 3.3 Dynamics of a GPT in Case (i)

impact and the economy returns to the original (steady state) growth rate.

Case (ii) An ideas generating GPT

In this case, f and Q enter multiplicatively in the knowledge pool, (curve

(3.4M)). An unanticipated new GPT arrives at time 0 when the economy is in the steady state (the steady state with a α=0.6, φ = 1.05 and z=0.1). The level of the GPT is increased by 15 per cent from 0.1 to 0.115. In Figure 3.4 the horizontal lines indicate the initial steady state whereas the curved lines indicate the transition to a new steady state. It is easily seen that a more sophisticated GPT implies a lower equilibrium knowledge gap, a higher steady state growth rate and a lower relative wage in the long run. On impact, the state variable R is, by definition, not affected. The new GPT, however, enlarges the knowledge pool and makes it attractive to invest in assimilation, therefore u jumps down (panel (b)). That is, the allocation of skilled labour immediately jumps towards assimilation activities at the 'cost' of accumulating productive knowledge. Therefore with a given ratio of productive knowledge to accumulation capabilities less research activity implies a lower growth rate of productivity in final goods production. Therefore the growth rate of GDP is lower initially (panel (c)).

Figure 3.5 presents the results of an equivalent shock given to the steady state shown in Table 3.3 with the same z and α=0.7 (here φ<1). The qualitative dynamics depicted are similar to the dynamics depicted in Figure 3.4 as far as panels (a), (b) and (c) are concerned. A discussion of panel (d) is postponed.

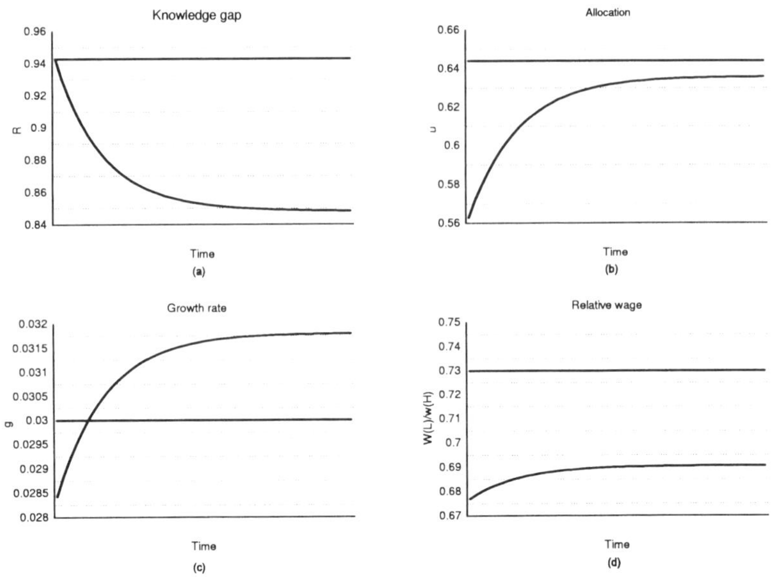

Figure 3.4 Dynamics of a GPT in Case (ii) with fishing out, $\varphi > 1$

Summarizing the results so far we can conclude the following. Independent of the question of whether the emergence of a new GPT will alter the long run growth rate the emergence a widely applicable technology as such will lead to stagnating productivity growth at impact.

How do the results match with the discussion on wage inequality? One lesson is that the question posed by Krugman (1995b): 'has the growth in total factor productivity been sufficient to be consistent with the large changes we have actually seen in factor prices?' (p. 7, 8) can be answered positively.

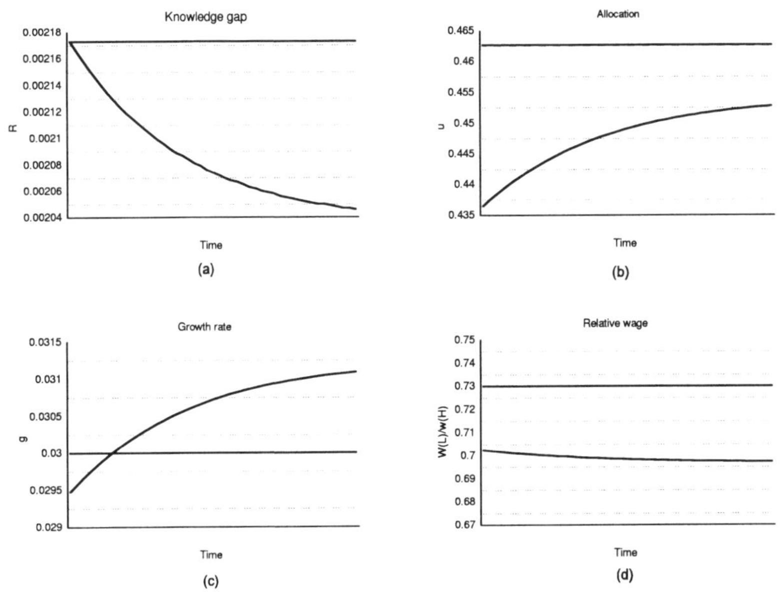

Figure 3.5 Dynamics of a GPT: Case (ii); intertemporal learning, $\varphi < 1$

The lower growth rate is even 'necessary' to increase wage inequality. A point that should be made before discussing the actual results is that empirical research is required to determine the magnitude of the shock to the GPT that we have modelled here. For now such information is lacking. Comparing Figure 3.4 to Figure 3.5 shows that the qualitative dynamics differ between the two sets of parameters we have used. In the simulation with fishing out, Figure 3.4, the bad news for unskilled workers is over: at impact the relative wage jumps down whereas during the transition the relative position of unskilled workers improves. In case of an ideas generating GPT with a positive learning effect in assimilation the relative wage of unskilled workers decreases further (see Figure 3.5). The intuition for the difference is that the return to assimilation declines quickly due to assimilation, if fishing out is relevant, see equation (3.10). In that case the ideas generating impact of the GPT is counterbalanced by a negative learning effect causing research, u, to increase much faster than in the case with intertemporal learning (Figure 3.5).

How do our results match quantitatively? Davis (1992) reports a ratio of unskilled relative to skilled worker wages of 0.66 for 1987 (compared to 0.73 in 1979). The simulations yield a wage ratio that is close to this number. The results are especially important in the light of the puzzle raised by Feenstra and Hanson 'the large increase in the non-production wage share over the period 1979–1987 is primarily the result of an increase in the relative employment of non-production workers that *occurred in just two years*, 1979 and 1980' (italics added, 1996 p. 8).[39]

7. FINAL REMARKS

Two empirical puzzles that emerged in the 1980s are the increase in the skill premium and sluggish productivity growth despite a technological revolution. Previous literature deals with these empirical phenomena separately. This chapter shows a natural way to integrate both. It has been shown that a superior General Purpose Technology leads to a temporary slump in the growth rate and an increase in the skill premium at the same time.

Most importantly, this model overcomes the critique of earlier work on GPTs (especially HT's) that requires reallocation from research to production and vice versa. This model generates a cycle without reallocation of labour from production to research. Secondly, the model generates these cycles without necessarily assuming that the new GPT is incompatible with existing knowledge stocks.

NOTES

1. Computer and information technology are used throughout this book interchangeably to indicate information processing equipment, software and applications of microprocessors in general.

2. Many observers argue that in the late 1990s the paradox evaporated, as productivity growth accelerated. This is exactly what this chapter predicts. Though, admittedly, the acceleration in productivity growth is largely driven by the computer-producing sectors (see Chapter 1).
3. Chapter 6, however, shows that this intuition is not necessarily correct.
4. See Chapter 8 for an analysis that assesses the maximum impact of trade.
5. See Krueger (1993) for evidence on the wage premium due to computer use. For a more thorough and extensive discussion of the potential explanations for the increase in wage inequality, see Chapter 2.
6. In most endogenous growth models a single R&D activity generates blueprints, see for example Grossman and Helpman (1991).
7. These workers are called R&D workers in the remainder but you might want to think of these in a more broad sense as all workers who experience substantial learning possibilities in their jobs, including for example most white-collar workers (Chapter 6 is more elaborate on this perspective).
8. At that time, as today, measurement problems might have played a role. The replacement of gas lighting by electrical lighting improved brightness, safety etc. without directly affecting recorded productivity.
9. Hence, if anything, recorded labour productivity might have increased, not TFP, as the capital-labour ratio increased due to a 'double' capital stock.
10. Bresnahan and Trajtenberg (1995) cite Griliches' study of hybrid corn. Hybrid corn is a technology that generated completely new possibilities in the field of agriculture: 'Hybrid corn was the invention of a method of inventing, a method of breeding superior corn' (p. 501, 1957).
11. Rustichini and Schmitz (1991) use such a structure to analyse optimal technology policy. The structure of this learning process resembles the idea of learning to learn, see Stiglitz (1987).
12. An extension of the model where skilled workers are specialized in research and unskilled workers in production allows analysis of the skill premium over the cycle. In the first phase, the skill premium increases, in the second phase a non-monotonic pattern prevents derivation of unambiguous results. Unattractive features of this version of the model are the decline of real wages of skilled workers in the second phase and the fact that there is no allocation decision what-so-ever; skilled workers produce new components and unskilled workers produce final goods.
13. Eriksson and Lindh (1997) endogenize the arrival rate of GPTs and allow for intertemporal spillovers in the HT framework. Helpman and Trajtenberg (1996) analyse the diffusion of GPTs throughout the economy over heterogeneous final goods.
14. The IT revolution thus induces an increased renewal of plants. These plants all need to master the technology, yielding a temporary slowdown in growth and an increased skill premium as implementation and learning are skill intensive.

15. The adoption of the diesel locomotive in the US was not associated with an initial slump in output. Other factors in the model that counteract a decline in output are the price decline on the second-hand market for capital goods and the more intense use of old capital. Both might have been relevant for the example.
16. Cheng and Dinopoulos generate growth cycles by an R&D sector that can target either on breakthroughs or improvements. If the return to improvements does not quickly diminish a breakthrough will be followed by a sequence of improvements, hence generating a cyclical pattern. See also Jovanovic and Rob (1990) on extensive and intensive search. Stein (1997) models two dimensions along which technology improves. Incumbent firms increase their lead over potential entrants in one technological dimension by learning-by-doing. Once this lead is substantial, only very favourable circumstances induce potential entrants to enter, who, by doing so, make future entry much more likely (this mechanism generates an uneven growth pattern).
17. Galor and Moav (1998) focus on within-group wage inequality.
18. This point was first made by Aghion and Howitt (1996).
19. Time subscripts are not introduced where it leads to no confusion.
20. This can be motivated by assuming that the effective application of knowledge is only possible if a firm has developed knowledge by its own R&D. Alternatively, patents could take care of monopolizing the knowledge stock.
21. For most readers the most nearby example, to make the discussion less abstract, is doing research while only reading own previous work. It will be clear that in the end the effective additions to your knowledge stop.
22. Again take the academic example: if you are working on a paper with three co-authors adding another four authors will not double output.
23. To continue the academic analogy: the output of writing summaries of books and papers is doubled when the input is doubled.
24. The number of firms is not an explicit argument in Z as it is fixed throughout the analysis. An alternative motivation could be that all firms have access to an unweighted average knowledge stock and hence N is irrelevant. De Groot and Nahuis (1998) analyse the interaction between N and f in a much simpler framework. Peretto and Smulders (2002) analyse the interaction between spillovers and changes in the number of firms.
25. This is easily seen by combining equations (3.1) and (3.15).
26. For the analysis of returns to variety see Chapter 7.
27. The expression can also be read as $\hat{f} = \hat{k}$ where a hat denotes a growth rate, hence: $\hat{f} \equiv \frac{\partial f / \partial t}{f}$.
28. The parameter restriction is implied by $u > u^{\sharp}$. $u^{\sharp} = (\varphi + (\rho - 1))/(\varphi + (\rho - 1) + \alpha/(1-\alpha))$ hence if α close to unity $u^{\sharp}$ collapses to the vertical axis. If α goes to zero $u^{\sharp}$ goes to unity, hence the equilibrium u is close to one therefore growth will be driven almost completely by research.
29. Technically speaking, for the solution of the growth rate equation (3.9) is not required. After solving for u and R the relative wage can be solved for recursively.
30. The comparative statics results are derived under a logarithmic utility function.

31. A third case is briefly discussed in Appendix E.
32. Divide both sides of (3.4A) by f, due to (3.3), and notice that the impact of Q vanishes.
33. Appendix B explains the derivation of Y and discusses other comparative statics exercises.
34. Dinopoulos and Thompson (1995, 1996).
35. In case (i) a second state variable prevails, see Appendix D for computational details. In this case the numerical examples have two positive and two negative roots.
36. The simulation shown in Figure 3.3 is based on the calibration exercise α=0.7 and z=1 as predetermined parameters.
37. Note that in the transition R is not an appropriate indicator of the knowledge gap.
38. If GDP also registered intangible investments the slowdown in growth would not occur. Chapter 4 elaborates on measurement issues.
39. Of course in our model there is no reallocation of workers between production and non-production (research and assimilation) work but if the model allowed for substitution, the change in relative factor rewards would to some extent turn up in reallocation of workers.

APPENDICES

A. The linearization procedure

The linearization procedure of the model is somewhat complicated by the fact that some variables are stationary and others are non-stationary. This appendix explains the procedure (and draws heavily on Smulders, 1994). The linearization procedure for static equations is standard. By taking total differentials and dividing by the initial value of a variable we obtain percentage deviations from the original steady state.

$$dw \equiv d^{new} - d^{old} \ . \tag{A.1}$$

To simplify notation we define variables with a tilde as:

$$\tilde{w} \equiv \frac{dw}{w^{old}} \ . \tag{A.2}$$

Where it leads to no confusion the superscripts for old and new (steady) states are dropped. A simple example is the mark-up relation,

$$\tilde{p} = \tilde{w}_l - \tilde{f} = 0 \ . \tag{A.3}$$

As f is non-stationary the level is not determined. However, as the growth rate is stationary, the change in the growth rate can be determined. So if $\dot{x}/x = g_x$ the growth rate of a *non-stationary* variable reads in linearized form:

$$d(\dot{x}/x) = g\tilde{g}_x \qquad \frac{d(\dot{x}/x)}{\dot{x}/x} = \tilde{g}_x \ . \tag{A.4}$$

Growth rates of stationary variables (examples are q_h, q_f but also u; required for the log-time differentiated version of the optimality condition (3.7)) can be derived as follows:

$$d(\dot{y}/y) = \frac{1}{y}d\dot{y} - \frac{\dot{y}}{y^2}dy = \frac{1}{y}d\dot{y} \quad , \tag{A.5}$$

where we use the fact that as we linearize a stationary variable around the steady state these variables grow at rate zero. By definition:

$$d\dot{y} \equiv \dot{y}^{new} - \dot{y}^{old} \quad , \tag{A.6}$$

and $\dot{y}^{old} \equiv 0$. To derive $\dot{y}^{new}/y$, the following steps are helpful:

$$dy \equiv y^{new} - y^{old} \quad , \tag{A.7}$$

is equivalent to:

$$y^{new} = y^{old}\left(1 + \frac{dy}{y^{old}}\right) = y^{old}(1 + \tilde{y}) \quad . \tag{A.8}$$

Differentiating with respect to time yields

$$\dot{y}^{new} = \dot{y}^{old}\tilde{y} + \dot{\tilde{y}}y^{old} \quad . \tag{A.9}$$

And use again that $\dot{y}^{old}$ is 0 by definition; hence we derive:

$$\frac{\dot{y}^{new}}{y^{old}} = \dot{\tilde{y}} \quad . \tag{A.10}$$

So :

$$d(\dot{y}/y) = \dot{\tilde{y}} \ . \tag{A.11}$$

Note that in the new steady state $\dot{\tilde{y}}=0$.

Finally, z is the ratio of two non-stationary variables. As all non-stationary variables grow at a common rate this ratio itself is a stationary variable. Define $z=x_1/x_2$. The rate of change of the stationary variable is difference between two growth rates of the non-stationary variables:

$$d(\dot{z}/z)=d\left(\frac{\dot{\overline{x_1/x_2}}}{x_1/x_2}\right)=g\tilde{g}_{x_1}-g\tilde{g}_{x_2} \quad . \qquad \text{(A.12)}$$

In the steady state all non-stationary variables grow at a common constant rate hence $\dot{\tilde{z}}=0$.

B. Steady-state analysis

B.1 Steady-state ratios

To solve for the asset portfolio in the steady state use (3.7) and (3.8) and substitute the two accumulation equations (3.2), (3.3):

$$\frac{q_f f}{q_k k}=\frac{u}{1-u}\frac{1}{1-\alpha} \quad . \qquad \text{(B.1)}$$

From (3.10) is alternatively solved for the asset portfolio by substituting (3.2) and (3.3), hence:

$$\frac{q_f f}{q_k k}=\frac{1}{\alpha}\left[\left(\frac{r}{g}-1\right)+\varphi\right] \quad . \qquad \text{(B.2)}$$

The transversality condition learns that $r/g>1$. Combining the previous two equations yields:

$$\frac{u}{1-u}=\frac{1-\alpha}{\alpha}\left[\left(\frac{r}{g}-1\right)+\varphi\right] . \tag{B.3}$$

Where the l.h.s. is the ratio of the wage bill of workers engaged in research versus those in assimilation. It is convenient for later reference to derive the following expression from the no-arbitrage condition for research (3.9) (use (3.7)):

$$\omega\frac{L_N}{H_N}\frac{1}{u}=\frac{1}{1-\alpha}\left(\frac{r}{g}-1+\alpha\right) . \tag{B.4}$$

The LHS is the ratio of the wage bill of unskilled workers versus workers engaged in research.

B.2 Linearization around the steady state

Linearization of the Ramsey rule around the steady state yields:

$$\tilde{g}_x=\frac{r}{\rho g}\tilde{r}-\frac{\theta}{\rho g}\tilde{\theta} . \tag{B.5}$$

Equation (3.6) yields:

$$\tilde{w}_L=\tilde{f}, \qquad \tilde{g}_{w_L}=\tilde{g}_f . \tag{B.6}$$

The second expression is derived, as both variables are non-stationary (and divide both sides by g). Equation (3.7) yields:

$$\tilde{q}_f+\tilde{A}_f-\alpha\tilde{u}-\alpha\tilde{H}_N+(1-\alpha)\tilde{f}+\alpha\tilde{k}=\tilde{w}_H, \qquad \dot{\tilde{q}}_f-\alpha\dot{\tilde{u}}-\alpha g\tilde{g}_f+\alpha g\tilde{g}_k=g\tilde{g}_{w_H}-g\tilde{g}_f . \tag{B.7}$$

Equation (3.8) yields:

$$\tilde{q}_k+\varphi\tilde{z}+\varphi\tilde{f}+(1-\varphi)\tilde{k}=\tilde{w}_H, \qquad \dot{\tilde{q}}_k+\varphi g\tilde{g}_f-\varphi g\tilde{g}_k=g\tilde{g}_{w_H}-g\tilde{g}_k \quad . \tag{B.8}$$

The no-arbitrage condition for knowledge accumulation, (3.9) and substituting (3.7) in there and linearizing yields:

$$\dot{\tilde{q}}_f= r\tilde{r} - r\left[\tilde{A}_f-\alpha(\tilde{f}-\tilde{k})+\frac{(1-\alpha)uH_N-\alpha\omega L_N}{\omega L_N+uH_N}(\tilde{u}+\tilde{H}_N)+\frac{\omega L_N}{\omega L_N+uH_N}(\tilde{L}_N+\tilde{w}_L+\tilde{w}_H)\right], \tag{B.9}$$

where the weights can be simplified by dividing numerator and denominator by uH_N and substituting for (B.4). This yields:

$$\dot{\tilde{q}}_f= r\tilde{r}- r\left[\tilde{A}_f-\alpha(\tilde{f}-\tilde{k})+\left((1-\alpha)\frac{g}{r}-\alpha\right)(\tilde{u}+\tilde{H}_N)+\left(1-(1-\alpha)\frac{g}{r}\right)(\tilde{L}_N+\tilde{w}_L+\tilde{w}_H)\right]. \tag{B.10}$$

The no-arbitrage condition for assimilation (3.10) can be rewritten (use (3.7) and (3.8)) as:

$$A_f(zf)^{\varphi}k^{-\varphi}H_N\left(\frac{\alpha}{1-\alpha}u+(1-\varphi)(1-u)\right)+\frac{\dot{q}_k}{q_k}=r \quad . \tag{B.11}$$

Linearizing yields:

$$\dot{\tilde{q}}_k=r\tilde{r}-r\left[\tilde{A}_k+\varphi\tilde{z}+\varphi(\tilde{f}-\tilde{k})+\frac{\frac{\alpha}{1-\alpha}u-(1-\varphi)u}{\frac{\alpha}{1-\alpha}u+(1-\varphi)(1-u)}\tilde{u}+\tilde{H}_N\right] \quad . \tag{B.12}$$

The weight is again expressed in terms of r and g by substituting (B.3):

$$\dot{\tilde{q}}_k = r\tilde{r} - r\left[\tilde{A}_k + \varphi\tilde{z} + \varphi(\tilde{f} - \tilde{k}) + \left(1 - \frac{g}{r}(1-\varphi)\right)\left(1 - \frac{(1-\varphi)(1-\alpha)}{\alpha}\right)\tilde{u} + \tilde{H}_N\right] \quad . \qquad \text{(B.13)}$$

And finally the two accumulation equations (3.2), (3.3) yield:

$$\tilde{g}_f = \tilde{A}_f + (1-\alpha)\tilde{u} + (1-\alpha)\tilde{H}_N - \alpha(\tilde{f} - \tilde{k}) \quad , \qquad \text{(B.14)}$$

$$\tilde{g}_k = \tilde{A}_k + \varphi\tilde{z} + \varphi(\tilde{f} - \tilde{k}) - \frac{1-\alpha}{\alpha}\left(\frac{r}{g} - (1-\varphi)\right)\tilde{u} + \tilde{H}_N \quad . \qquad \text{(B.15)}$$

where (B.3) is substituted.

The steady-state definition in (3.19) and (3.20) reads in linearized form:

$$\tilde{g}_x = \tilde{g}_f = \tilde{g}_k = \tilde{g}_{w_L} = \tilde{g}_{w_H} \equiv \tilde{g}, \qquad \dot{\tilde{q}}_f = \dot{\tilde{q}}_k = 0 \quad . \qquad \text{(B.16)}$$

Finally the constant allocation in the steady state implies:

$$\dot{\tilde{u}} = 0 \quad . \qquad \text{(B.17)}$$

B.3 An improved GPT: an example

To find the impact of an increased quality of the GPT on the steady state substitute the steady-state definitions (B.16) and (B.17) in equations (B.5)-(B.15). And set:

$$\tilde{\theta}=\tilde{A}_f=\tilde{A}_k=\tilde{H}_N=\tilde{L}_N=0 \quad . \tag{B.18}$$

To solve for $\tilde{g}$ as a function of $\tilde{z}$ substitute (B.5) in (B.13) and plug in $(\tilde{f}-\tilde{k})$. $(\tilde{f}-\tilde{k})$ follows from (B.14)=(B.15). This yields $\tilde{g}$ is a function of $\tilde{u}$. Plug $(\tilde{f}-\tilde{k})$ in (B.14) to get a second expression for $\tilde{g}$ in terms of $\tilde{u}$. Solving for $\tilde{g}$ and simplifying yields Y in (3.23).

B.4 Comparative statics

Table B.1 shows the comparative statics results, derived with the linearized model. The parameters capturing the efficiency of the growth engines both have a positive impact on the growth rate, although the mechanism differs. An increase in the efficiency parameter of assimilation, A_k, increases the return to assimilation and hence induces a reallocation of skilled workers from research to assimilation. Obviously, the knowledge gap decreases as the efficiency of, and resources allocated to, assimilation increase. An increased efficiency of research, A_f, leads *ceteris paribus* to an increased return to research and an increase in the growth rate of productive knowledge. This would induce reallocation of workers towards research, where account has to be taken of the fact that decreasing returns to research labour mitigates this effect. Productive knowledge spills over to the knowledge pool and hence increases the return to assimilation. This induces reallocation towards assimilation. On balance more resources are allocated to assimilation, but the knowledge gap decreases. Obviously, marginal increases in the productivity of skilled workers lead to increased inequality.

Increasing the number of skilled workers increases the rate of growth. This result is the scale effect that prevails in many growth models. A larger economy, which should be interpreted as more skilled workers per firm, generates a higher growth rate. An increase in the amount of skilled workers leads to a decrease in the share of workers doing research, *u,* due to decreasing returns in research. As a larger chunk of the skilled workforce assimilates, the knowledge gap decreases.

Table B.1 Comparative statics

	θ	A_k	A_f	H_N	L_N
g	-	+	+	+	0
w_l/w_h	+	-	-	?	-
R	+	-	+	-	0
u	+	-	-	-	0

At first sight counter intuitively, an increase in the amount of skilled workers does not necessarily imply an increase in the relative wage of unskilled workers. An increase in the number of skilled workers makes them relatively abundant, and keeping everything else constant this would cause downward pressure on wages of skilled workers, but this need not be so. Keep in mind that skilled workers are only active in research and assimilation. So, to get a decline in the relative wage of unskilled workers the growth engine should become marginally more efficient as more workers are employed. It is necessary for this odd result to hold that $\varphi < 1$, hence the case where increased accumulation capabilities ease future assimilation (if φ is high the best ideas are fished out first and increased assimilation activity is not a very efficient activity).[1] The intuition for the odd result goes as follows. The share of workers in assimilation increases[2] and with very low φ the activity remains very efficient and does not rely very much on the general knowledge pool. Hence, the role for spillovers is small. The latter can also be understood from the other side: the fact that relatively fewer spillovers are generated due to the reallocation of research towards assimilation implies that spillovers should not be important, otherwise the growth engine could not be more efficient.[3] See Chapter 6 for an elaboration on the issue why spillovers should be small.

C. Corner solutions

C.1 Binding inequality constraints

In the assimilation technology (3.3) there are no Inada conditions, therefore at a certain rate of imbalance ($f/k \neq (f/k)^{\text{steady state}}$) one of the inequality constraints in Table 3.1 will become binding. Suppose the ratio of productive knowledge to accumulation capabilities is too low for further investment in accumulation capabilities to be attractive. Hence the condition $H_A \geq 0$ is binding.

The firm problem then becomes:

$$\max \; V_i = \int_0^{\infty} e^{-rt} \left(x_{it} p_{it} - H_{Rit} w_{Ht} - L_{it} w_{Lt} \right) dt \quad , \tag{C.1}$$

subject to: (3.1), (3.2), and (3.15). This yields (3.6),(3.7), and (3.9). The labour market condition for skilled workers now becomes

$$\frac{H}{N} = H_{Ri} \equiv H_N \quad . \tag{C.2}$$

Now the growth rate is easily determined as:

$$\hat{f} = A_f k^{\alpha} (H_N)^{1-\alpha} f^{-\alpha} \quad . \tag{C.3}$$

Having an initial value for f_0, the growth rate is determined; recall that k is constant. The relative change in the growth rate is $-\alpha\hat{f}$. Hence, the rate of growth is increasing in the imbalance and decreases after the shock. Use (3.7) and take w_H as a numeraire and log-differentiate to see:

$$\hat{q}_f = -(1-\alpha)\hat{f} \quad . \tag{C.4}$$

Hence, the shadow price of productive knowledge capital declines with the growth of productive knowledge. At a certain f/k ratio the inequality constraint ceases to be binding and the economy enters the regime discussed in the main text.

C.2 $\lambda<1$ and Z homogeneous of degree one in f

The approach proposed by Jones (1995b), $\lambda<1$, yields a model with undesirable features given the segmented labour market. Insert the definition of u in (3.2) and (3.3), divide respectively by f and k and log-differentiate to time, to get:

$$\hat{g}_f=(1-\alpha)\hat{u}-\alpha\hat{f}+\alpha\hat{k} \quad , \tag{C.5}$$

$$\hat{g}_k=\varphi\lambda\hat{Z}+(\lambda-1-\varphi\lambda)\hat{k}-\frac{u}{1-u}\hat{u}+\hat{f} \quad . \tag{C.6}$$

To have a steady state with constant growth rates and allocation requires:

$$\hat{f}=\hat{k} \quad \wedge \quad (\lambda\varphi+1-\lambda)\hat{k}=\lambda\varphi\hat{Z} \quad , \tag{C.7}$$

which implies:

$$\hat{Z}=\frac{\lambda\varphi-\lambda+1}{\lambda\varphi}\hat{f} \quad \vee \quad \hat{f}=\hat{k}=0 \quad . \tag{C.8}$$

In the main text (Section 4.1) we ruled out the first equality, as the parameter combination on the r.h.s. exceeds one. The second equality is hence needed to have a steady state. Hence a steady state with growth rate of zero would result. The allocation of skilled workers in no longer determined in that case.

Suppose skilled workers could be allocated symmetrically to firms to do R&D, a positive growth rate could be generated by a positive rate of growth in the endowment of skilled workers. No steady allocation could be reached as u is affected by changes in H, see Appendix B.

D. Dynamics of a GPT with the (3.4A) curve

The dynamic analysis of a GPT that enters the knowledge pool additively complicates the derivation of the differential equations describing the economy's behaviour outside the steady state somewhat. Taking the (3.4A) curve, we need to follow the same procedure as in Section 6.1. Taking into account that logarithmic differentiation to time of Z yields:

$$\hat{Z}=\frac{\dot{f}}{f+Q}=\frac{A_f R^{-\alpha}(uH)^{1-\alpha}}{1+\frac{Q}{k}\frac{1}{R}}\ , \tag{D.1}$$

where the last equality follows from (3.2) and the definition of R. Using this and the fact that $Z=f+Q$ the differential equation for u can be written as:

$$\begin{aligned}\dot{u}= &\frac{u}{\alpha}\left(\left(\frac{\alpha}{1-\alpha}u+\alpha(1-u)\right)A_k H_N\left(R+\frac{Q}{k}\right)^{\varphi}-(1-\alpha)\omega L_N A_f(uH_N)^{-\alpha}R^{-\alpha}\right)+\\ &\frac{u}{\alpha}\left(-\varphi\left(\frac{A_f(uH_N)^{1-\alpha}R^{-\alpha}}{1+\frac{Q}{k}\frac{1}{R}}\right)\right)\ .\end{aligned} \tag{D.2}$$

The equation for the ratio of productive knowledge to accumulation capabilities now looks like:

$$\dot{R}=R\left(A_f(uH_N)^{1-\alpha}R^{-\alpha}-A_k\left(R+\frac{Q}{k}\right)^{\varphi}(1-u)H_N\right)\ . \tag{D.3}$$

The dynamics of relative wages are determined by:

$$\dot{\omega}=\omega\left((1-\varphi-\rho)\left(\frac{A_f(uH_N)^{1-\alpha}R^{-\alpha}}{1+\frac{Q}{k}\frac{1}{R}}\right)-\theta+\frac{\alpha}{1-\alpha}(uH_N)A_k\left(R+\frac{Q}{k}\right)^{\varphi}\right) . \tag{D.4}$$

Finally the differential equation for Q/k is:

$$\frac{\dot{Q}}{k}=-\frac{Q}{k}A_k\left(R+\frac{Q}{k}\right)^{\varphi}(1-u)H_N . \tag{D.5}$$

Note that k grows over time and Q is constant, hence that in the limit the system of differential equations evolves to the one in the main text, as the last differential equation tends to 0=0.

E. A destructive GPT, case (iii)

Case (iii) is characterized by creative destruction. Hence there is negative interaction of the new GPT with productive knowledge already developed at time t applying the old GPT, hence $Z_{fQ}<0$. The radically new GPT makes existing knowledge redundant. The impact of a GPT that arrives at time t on the knowledge stocks that enter accumulation equation (3.3) can be formalized as follows:

$$Z_t=\bar{f}_t Q_j , \tag{E.1}$$

$$\bar{f}_t = (1-\delta_k)\bar{f}_s + \int_s^t \int_o^N \frac{1}{n}\dot{f}_{i\tau}\, dn\, d\tau \ . \tag{E.2}$$

The factor (1-δ) is the rate of complementarity between GPT *j-1* and GPT *j*. To be consistent, the experience that enters (3.2) is depreciated analogously but possibly with a different rate:

$$f_t = (1-\delta_f) f_s + \int_s^t \dot{f}_\tau\, d\tau \ . \tag{E.3}$$

Labour productivity l should not be depreciated. It is easy to see that, once adoption occurred, the long-run impact on growth is equivalent to Case (ii) as the magnitude of the redundancy of knowledge through creative destruction vanishes in the limit. Future research should provide an answer to the adoption decision. Here we abstain from the adoption issue. Once GPT j arrives, GPT *j-1* is no longer available by assumption. Hence, the destruction of creations with the previous GPT has only level effects.

NOTES

1. The exact condition is not very informative.
2. From equation (3.22) in the main text it is seen that the relative wage of unskilled workers is increasing in u.
3. The role for α in the condition is less clear, as an increase in α implies stronger diminishing returns to skilled workers in research at a point in time but also increases the share of accumulation capabilities in the research engine.

4. We don't see what we learn: The Solow Residual, a GPT and inequality

1. INTRODUCTION

Increased wage inequality between the high and low skilled has been widely observed in industrialized countries in recent decades. Davis (1992) shows that wage inequality increased in the 1980s in almost all OECD countries. In this chapter we argue that biased technological change associated with the arrival of a GPT is the major reason for increased wage inequality. Two other reasons are often given. First, lower educational attainment and hence increased scarcity of the highly educated tends to increase wage inequality. However, educational attainment has not been decreasing: there is a stable upward trend in the relative supply of skilled workers. Consequently, increased wage inequality is hard to explain by changes in education.[1] Second, trade with low-wage countries tends to cause downward pressure on wages of unskilled workers, thereby increasing inequality. This argument is met with scepticism as trade with low-wage countries is only a small component of total trade in the industrialized world (see Chapter 2 for extensive discussion of these reasons).

We think that today's GPT, the 'computer revolution', is a major contributor to the increase in wage inequality. A number of points suggests that the technology- related explanation is in line with increases in wage inequality. First, the introduction of information and communication technology (further ICT) is argued to be associated directly with a rising skill premium, see, for example, Krueger (1993).[2] Second, the increase in wage inequality parallels a significant increase in ICT investments. Allen (1997) provides data on ICT investments that show an acceleration in ICT investments in the early 1980s. Feenstra and Hanson (1996) show that inequality increased sharply in the period 1979–83. This timing of events is suggestive. Third, the US seems to be experiencing a *slowdown* in relative demand for skilled workers in the most recent years.[3] This empirical fact is hard to explain by trade with low-wage countries, as the relative demand for high-skilled workers is decreasing again where trade with low-wage countries is still increasing.[4]

Critical in the technology explanation is that the 'computer revolution' should accelerate the rate of technological change, or at least the rate of skill-biased technological change.[5] The problem is that we do not observe that measured technological change has increased since the introduction of ICT in the early 1980s. This puzzling observation has been expressed by Solow (1987) as: 'You can see the computer age everywhere but in the productivity statistics'. Put differently, the new GPT did not raise the first moment of the wage distribution, nor did it show up in the so-called Solow residual.

A number of explanations for this seemingly paradoxical finding have been put forward. Oliner and Sichel (1994) claim that the stock of ICT capital is still so small that it obviously has a negligible influence on productivity. Another explanation for the paradox is that productivity increases are difficult to measure because it is hard to capture quality improvements (due to increased use of computers) which are not reflected by higher prices (see Griliches, 1994). Griliches argues that an ever increasing part of the economy is 'unmeasurable'. Developing the argument that the rate and skill bias of technological change are related somewhat further, one could argue that, if the 'computer revolution' did not affect productivity growth, the hypothesis that increased inequality is a result of biased technical change is likely to be invalid.

The first point this chapter makes is that, contrary to the common sense argument, a computer-related bias in technological change is likely to *cause* a (temporary) slowdown in measured output growth. Moreover, the computer related GPT increases inequality upon arrival. This implies that the first and second moment of the wage distribution are related in a different manner from what one expects at first sight. Hence, we show that lower output growth and a bias in technology are naturally related.[6] As such, biased technological change can be a main contributor to increased wage inequality.

The second contribution of this chapter is to provide a solution to the 'productivity paradox'. The empirical puzzle that we do not observe the computer revolution in the growth statistics can be resolved by accounting for growth appropriately. We conjecture that in current productivity measures the effective labour inputs are incorrectly measured. As a consequence of the arrival of the GPT, skilled workers invest some of their time to master the new technology, in order to become more productive later on. This implies that if one does not correct for this decrease in time spent producing, one overstates the role of labour inputs, and one understates the role of technical change. A simple calculation reveals that an error of 0.2 percentage points of annual TFP growth is to be attributed to the increased learning time. We show that 33 per cent of the slowdown in productivity growth during the 1980s is illusionary.

This chapter also contributes to the literature on measurement in the 'knowledge economy', by explicitly spelling out the different fallacies in growth accounting. Finally, a dual approach to growth accounting is advocated in an economy where unobservable knowledge production is important, as the dual approach depends on observables only.

The rest of this chapter is organized as follows. Section 2, dealing with the first moment of the wage distribution, explains how and why technological change need not affect measured income. In Section 3 we develop an endogenous growth model of wage inequality and technological change to explore the relationship between the first and second moment of the wage distribution. In Section 4 we analyse the consequences of the computer revolution on wage inequality, growth, and technological change. Equipped with the mapping of the relations between the observable variables laid out in the model, we explain why we do not see the computer revolution in the productivity statistics. Next, it is shown how a dual approach to growth accounting can be used to resolve these measurement problems. This is done in Section 5. Section 6 concludes.

2. MEASUREMENT

The new GPT should affect the first moment of the wage distribution. The usual way to infer something about technological change is growth accounting. To explore how and why technological change might not turn out to affect average income, consider the following macroeconomic production function relating outputs to inputs (technology is assumed, D, to be Hicks-neutral):

$$\check{Y} = DG(K,L,H) \ , \tag{4.1}$$

where K , L, and H stand for the aggregate measures of capital, unskilled and skilled labour respectively. $\check{Y}$ denotes observed output. The Solow residual, or measured TFP, equals output growth that is not explained by changes in factor inputs. Time differentiating (4.1) and rearranging yields the fundamental relation between growth of output and growth of inputs:

$$\hat{D} = \hat{Y} - \chi_K\hat{K} - \chi_L\hat{L} - \chi_H\hat{H} \ , \tag{4.2}$$

where a hat denotes a relative change, and $\chi_i \equiv (D(\partial G/\partial i)i)/Y, \quad i = K,L,H$. Once the appropriate output and input concepts are defined, and when they are measured accurately, D is the technology indicator of interest. However, in a knowledge-based economy several pitfalls are to be taken into account to avoid that technological change does not turn up in the productivity statistics correctly. The remainder of this section explores potential pitfalls in defining output and input concepts.

First, what is the proper output concept? *Output*[7] of knowledge investments, for example human capital, due to on the job learning and R&D stocks, is not measured in national accounts. There is no obvious reason to treat additions to the knowledge capital stock any differently from those to the physical capital stock. Nevertheless, the production of capital is only measured if it concerns physical capital, but not if it is intangible knowledge capital. However, the production of both types of capital requires resources that otherwise could be used for current consumption. The difference is due to the fact that the item produced – knowledge capital – is not priced on the market, as the product usually is not sold or is unsaleable. The former holds for patents and the latter, for instance, for organizational capital. What would be appropriate is to include, alongside consumption and investment in physical capital, investment in knowledge capital to construct a broad output concept.

Second, the measurement accuracy of (narrow) output is problematic due to difficulties with incorporating quality improvements and new goods or varieties. This problem, however, is not specifically related to knowledge-based growth nor to a GPT and hence in the remainder it is assumed that output of goods and services is measured accurately.[8]

Third, the arrival of a new GPT might make existing capital obsolete. Hence, a technology shock could cause capital stocks to decay excessively rapidly. Thus, due to discontinuous depreciation, inputs might be measured inaccurately. If the national accounts were to account for knowledge as they do for physical capital this problem would even be more pressing. Aghion and Howitt (1998a) coin this the *obsolescence* problem.[9]

Finally, what is the proper input concept? This is a hard issue, but what is clear is that the input concept is related to the output concept (Section 5 explores this in greater detail). If, as the discussion of the *knowledge output* problem clarifies, produced knowledge is not taken into account, should not inputs be adjusted accordingly? Though it is true that successful knowledge creation raises GDP in the long run, resources should be adjusted for those who are accumulating knowledge. The remainder refers to this issue as the *knowledge input* problem.

Are the knowledge input and the knowledge output problem one and the same thing? And, how relevant are these factors in explaining a productivity paradox? Is it possible to relate measurement problems with respect to changes in the first moment of the wage distribution to those in the second moment? To answer these questions the next section sets out a minimalist endogenous growth model with two types of labour that relates the arrival of a GPT to observed productivity and wage inequality.

3. THE MODEL

We consider a closed economy with three decision units. First, there are firms which produce a unique product variety in a monopolistic environment. Second, there are competitive capital-assembly firms. Finally, households supply skilled (H) and unskilled (L) labour fully inelastically. This section subsequently explains the behaviour of these agents and solves for the model.

3.1 Production

Let there be one sector of production consisting of N firms indexed j, engaged in monopolistic competition. Every firm produces one product variety j facing a downward-sloping demand curve for that variety. We assume that there is no entry/exit, so that the number of firms is fixed.

Production in firm j of final goods (x_j) requires unskilled labour (L_j), skilled labour (H_j), capital goods (K_j), and technology-capital or firm-specific productive knowledge (f_j). Firms have to devote resources to research in order to remain competitive. Time that high-skilled workers are engaged in production is $u_j H_j$, where u_j is the fraction of the time endowment (which is normalized to unity) which is spent producing. The remainder of the time endowment is devoted to research (or learning). Unskilled workers are employed in production only. Production is designated by a Cobb-Douglas production function:[10]

$$x_j = Tf_j^{1-\alpha}K_j^{\alpha}\left[(u_jH_j)^{\beta}L_j^{1-\beta}\right]^{1-\alpha}, \tag{4.3}$$

where $\alpha,\beta \in (0,1)$ and T is the exogenously given technology level. This production function features the duplication argument: doubling physical inputs, K, uH, L, leads to a doubling of output. For later reference, we denote the productivity of physical inputs as $A \equiv Tf^{1-\alpha}$.

The production function is homogeneous of degree 1 in K and f. The reason is that, eventually, we want a steady-state solution in which u, H, and L are constant and K and f grow at the same rate. Each firm faces a downward-sloping demand curve for its variety:

$$x_j = (p_j/P_X)^{-\varepsilon}X \,, \tag{4.4}$$

where $\varepsilon > 1$ stands for the price elasticity of demand. P_X is the aggregate ideal price index and X is aggregate output (the demand curve is derived from demand from consumers and capital-assembly firms):

$$P_X \equiv \left(\frac{1}{N}\sum_{j=0}^{N} p_j^{\,1-\varepsilon}\right)^{\frac{1}{1-\varepsilon}}, \qquad X \equiv N\left(1/N\sum_{j=0}^{N} x_j^{\frac{\varepsilon-1}{\varepsilon}}\right)^{\frac{\varepsilon}{\varepsilon-1}} . \tag{4.5}$$

We focus on symmetric equilibria, hence prices and levels of production reduce to: $P_X=p_j=p$ and $X=x_j$ for all j as we normalize N to 1.

Firms maximize the net present value of profits, which is the discounted stream of instantaneous profit flows, π, at rate r: $\Pi_j=\int_0^\infty \pi_j \exp[-\int_0^t r(v)dv]dt$. Instantaneous profits are given by:

$$\pi_j = p_j x_j - rK_j - w_H H_j - w_L L_j \,, \tag{4.6}$$

where $w_H H$ and $w_L L$ are loan payments to skilled and unskilled workers. Capital K is hired for a rental rate of r from capital-assembly firms. In order to keep producing profitably, the firm devotes some of its high-skilled workers to the production of knowledge f:

$$\dot{f}_j = Q(i) f_j (1-u_j) H_j \,, \tag{4.7}$$

where a dot denotes a time derivative, thus $\dot{f} \equiv df/dt$ and Q denotes the productivity of research. The firm-specific nature of research implies that the productivity of a firm's research today depends on the accumulated stock of firm-specific knowledge capital.

This knowledge production function has two features attributable to Lucas (1988). First, as in Lucas (1988), accumulation of knowledge is intensive in human capital. No physical capital goods are inputs in production of knowledge.[11] Second, this chapter draws upon the notion of Lucas that it is *time* devoted to learning and researching that is important. In this interpretation not only the total amount of physical units of labour actually engaged in research but potentially also the amount of time 'non-researchers' spent on learning matters (an on-the-job-learning interpretation is suitable).

3.2 Capital assembly

Competitive capital-assembly firms assemble the N product varieties costlessly into capital. The production function for capital is:

$$I \equiv N\left(\frac{1}{N}\sum_{j=0}^{N} i_j^{\frac{\varepsilon-1}{\varepsilon}}\right)^{\frac{\varepsilon}{\varepsilon-1}} . \tag{4.8}$$

$P_I I$ are total purchases of capital goods (where P_I is the ideal price index). Investment demand for each variety is denoted i_j. Produced capital adds to the stock of capital (we abstain from depreciation):

$$I = \dot{K} . \tag{4.9}$$

The stock of capital generates a rent of rK paid by the monopolistic firms. This rent is paid as a dividend to consumers who supply the capital-assembly firms with funds.

3.3 Consumption

There is a representative household having a constant saving rate. Hence the consumption function looks like:

$$C = (1-s)pY , \tag{4.10}$$

where s is the saving rate[12] and Y is real income. Consumption of different varieties is derived from:

$$C \equiv N\left(\frac{1}{N}\sum_{j=0}^{N} c_j^{\frac{\varepsilon-1}{\varepsilon}}\right)^{\frac{\varepsilon}{\varepsilon-1}} . \tag{4.11}$$

The demand curves from the capital-assembly firms and consumers are identical and hence represented by a single downward-sloping demand curve for each variety (hence $x_j=c_j+i_j$). Goods-market equilibrium gives the resource constraint of the economy:

$$Y = C + I , \tag{4.12}$$

where it is used that $X=Y$.

3.4 Firms

Firms maximize profits subject to their demand function, accumulation constraints on firm-specific capital goods and the stock of knowledge. Solving the firm's optimal control problem by setting up the Hamiltonian we can derive the first-order conditions for maximization of profits (see Appendix A). In symmetric equilibrium:[13] $x_j=X$, $p_j=p$, $H_j=H$, $L_j=L$, $K_j=K$, $f_j=F$, and $u_j=u$. In the remainder we normalize prices: $p\equiv\varepsilon/(\varepsilon-1)$. The last normalization implies that the 'mark-up' equals the price of the variety. Some additional notation simplifies the exposition: $\Theta\equiv\beta(1-\alpha)$ and $\Xi\equiv(1-\beta)(1-\alpha)$.

First, the wage rate of the unskilled workers is equal to the marginal productivity:

$$w_L = \Xi T K^{\alpha} F^{1-\alpha} (uH)^{\Theta} L^{\Xi-1} . \tag{4.13}$$

Second, we can derive the skilled workers' wage and a no-arbitrage condition for the allocation of time of skilled workers in the production of goods and learning:

$$w_H = \Theta T K^{\alpha} F^{1-\alpha} (uH)^{\Theta-1} L^{\Xi} = q_f Q(i) F , \tag{4.14}$$

where q_f is the co-state variable associated with the accumulation constraint on technology. The marginal value of time skilled workers devote to production and research must be equal. The latter has to be multiplied by the relative price of technology in terms of final output, as measured by q_f.

The wage differential between unskilled and skilled workers, expressed as the ratio of the wages, is $\omega(u) \equiv w_L / w_H = (\Xi/\Theta)(H/L)u$. The wage differential increases (ω lower) as high-skilled workers spend more time learning and doing research (u lower): $\omega' > 0$. The dynamics of this economy are described by the evolution of time spent producing u, (see Appendix A):

$$\frac{\dot{u}}{u} = \left(\frac{1-\alpha}{\Theta}\right) Q(i) u H + \left(\frac{1-\alpha}{1-\Theta}\right) Q(i) H + \left(\frac{\alpha(1-sp)}{\Theta-1}\right) T R^{1-\alpha} (uH)^{\Theta} L^{\Xi} , \tag{4.15}$$

where R is the ratio of the two capital stocks ($R \equiv F/K$). The differential equation describing the evolution of R is given by (see Appendix A):

$$\frac{\dot{R}}{R} = Q(i)(1-u)H - spTR^{1-\alpha}(uH)^{\Theta} L^{\Xi} . \tag{4.16}$$

3.5 Steady state

A steady state is an equilibrium in which $\dot{R}=\dot{u}=0$. It is easy to show that the steady state is characterized by:

$$\frac{\dot{w}_H}{w_h} = \frac{\dot{w}_L}{w_L} = \frac{\dot{K}}{K} = \frac{\dot{F}}{F} \equiv g \ , \quad \frac{\dot{q}_f}{q_f} = \frac{\dot{p}_X}{p_X} \ . \tag{4.17}$$

We can derive the steady-state share of time spent producing (u^*) as:

$$u^* = \frac{\alpha - sp}{\alpha - \left(1 - \frac{1}{\beta}\right) sp} \ , \tag{4.18}$$

where $\alpha > sp$ is assumed in order to arrive at a feasible steady-state allocation ($0<u<1$).[14] Note that the steady-state value of u depends only on the production technology parameters α and β, the saving rate s and the inverse of the monopoly mark-up, p. Recall that $\omega(u) \equiv w_L/w_H = (\Xi/\Theta)(H/L)u$. Hence, the parameters in (4.18) and the relative supplies of high- and low-skilled workers determine wage inequality. Moreover, the technology-capital-stock ratio (R^*) is:

$$R^* = \left[\frac{Q(i)(1-u^*)H}{spT(u^*H)^\beta L^{1-\beta}}\right]^{\frac{1}{1-\alpha}} \ . \tag{4.19}$$

In Appendix A it is shown that this equilibrium is saddle-point stable under the condition: $u<(\Theta/(1-\Theta))^2$.[15] u is the 'jump' variable and R is the 'predetermined' variable.

The steady-state growth rate of the stocks (K and F), output (Y), and wages of skilled and unskilled workers (w_H, w_L) will be equal to:

$$g = Q(i)(1-u^*)H \ . \tag{4.20}$$

3.6 Transitional dynamics

The transition dynamics are simple. Figure 4.1 gives the phase portrait in the $R-u$ space. The $\dot{R}=0$ locus gives all combinations (R,u) where R is not changing. The $\dot{R}=0$ locus is downward sloping:

$$\left[\frac{du}{dR}\right]_{\dot{R}=0} = -\left(\frac{1-\alpha}{\frac{u}{1-u}+\Theta}\right)\frac{u}{R} < 0 \ . \tag{4.21}$$

The $\dot{u}=0$ locus gives, equivalently, all the points where u is not changing. Its slope is:

$$\left[\frac{du}{dR}\right]_{\dot{u}=0} = \left(\frac{(1-\alpha)\Phi\frac{1}{R}}{\left(\frac{1-\alpha}{\Theta}\right)Q(i)H-\frac{\Theta}{u}\Phi}\right) > 0 \ , \tag{4.22}$$

where $\Phi \equiv [(1-\alpha)/\Theta]Q(i)Hu+[(1-\alpha)/(1-\Theta)]Q(i)H$. The denominator is positive as a consequence of the stability condition, that is, when $u>(\Theta/(1-\Theta))^2$. The equilibrium can be found where the two phase lines intersect. Starting from an initial capital stock (K_0) and an initial firm-specific knowledge stock (F_0), a jump in u leads to the stable branch, and the economy converges along the stable branch towards the equilibrium.

4. A COMPUTER REVOLUTION...

A GPT, as the computer revolution, is likely to affect the marginal productivity of research as it opens up new opportunities for knowledge-creating activities throughout the economy. This chapter steps back from the detailed differences in GPTs explored in the previous chapter and simply mimics the computer revolution by an increase in Q. Accordingly $Q(i)$

increases to $Q(i+1)$. Hence, the productivity of learning or time spent doing research increases. Only the high-skilled workers can benefit initially from the increased productivity in learning. Time devoted to learning and assimilating the new technology becomes more valuable.

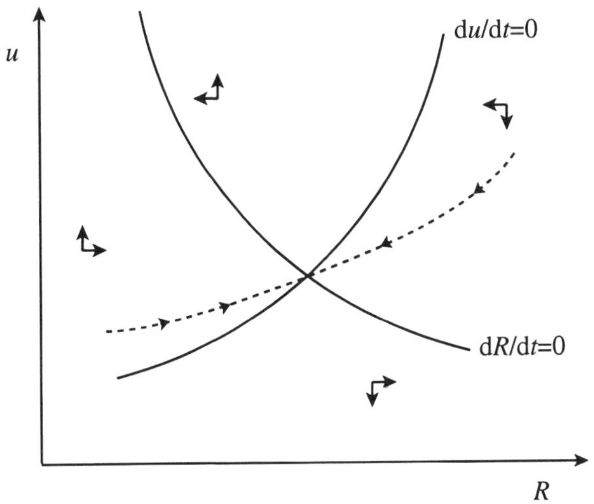

Figure 4.1 Phase diagram

Graphically, this can be represented by a rightward shift of both the $\dot{u}=0$ and the $\dot{R}=0$ loci; see Figure 4.2.

On impact u falls to equate the marginal productivity of high skilled workers in learning and producing. Therefore, more time will be spent learning

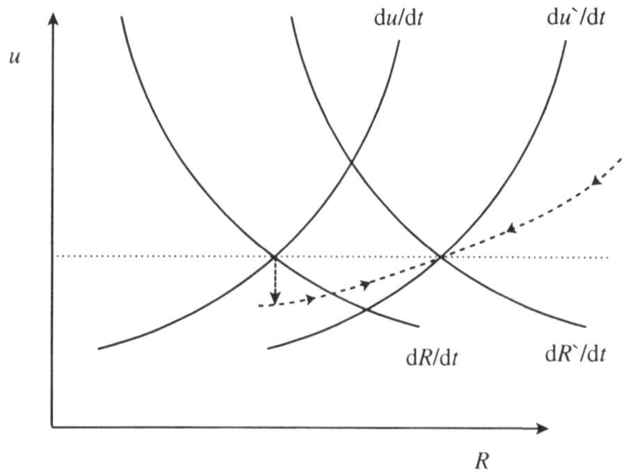

Figure 4.2 *Adjustment to the new steady state*

and less time will be spent producing in order to accumulate *F*. The growth rate of output falls, because *u* falls, see Figure 4.3(a). Over time output growth recovers as the stock of technology increases. Hence, high-skilled workers reduce time learning because a higher stock of knowledge increases the value of an hour producing; see also Figure 4.3(b). Wage inequality rises immediately after the shock; see Figure 4.3(c). The increase in *Q* leads to a permanently higher growth rate in the long run, see equation (4.20). The reason is that the productivity of an hour spent learning has increased permanently.

5. ... AND THE REASON WHY WE DON'T SEE IT

Measuring the total factor productivity effect of the computer revolution implies, as the modelling exercise above stresses, adjusting labour input time for learning and accounting for the production of firm-specific knowledge capital.

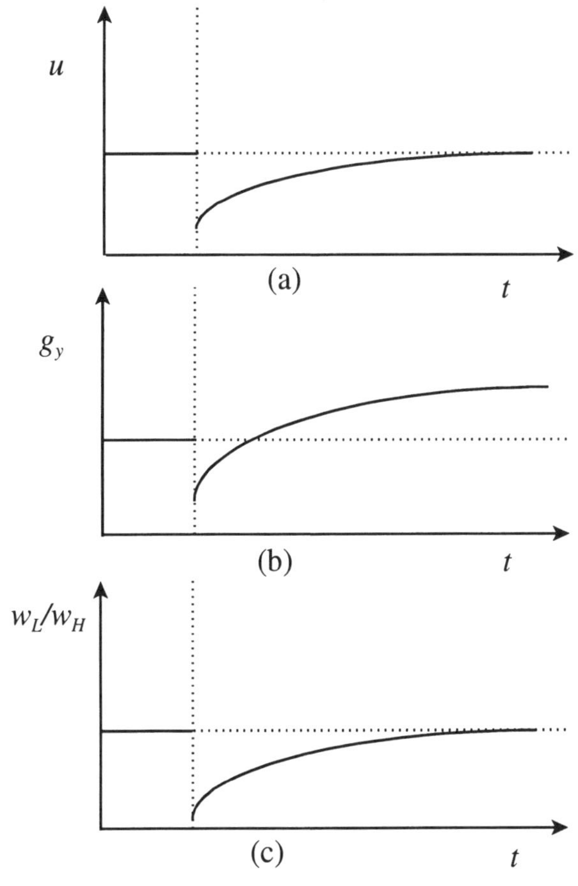

Figure 4.3 *Effects of an increase in Q*

5.1 The knowledge-input problem

Due to on-the-job learning, as we model the struggle with the new technology here, production falls. This fall in goods production is recorded in the national accounts, whereas the accompanying reduction in time spent producing due to on-the-job learning is not recorded.[16] This *knowledge input* problem tends to reduce measured TFP.[17]

To measure productivity appropriately, our model stresses that an adjustment for learning time should be made. Thus the production function looks like: $Y=AF(K,L,uH)$. Hence, to get actual TFP growth of physical production, inputs in physical production should be adjusted to the actual level:

$$\hat{A} = \hat{Y} - \chi_K\hat{K} - \chi_L\hat{L} - \breve{\chi}_H(\hat{H} + \hat{u}) , \qquad (4.23)$$

where $\breve{\chi}_H \equiv (A(\partial F/\partial uH)uH)/Y$.[18] The difference between *input* adjusted productivity growth, $\hat{A}$, and measured productivity growth, $\hat{D}$, is $-\breve{\chi}_H\hat{u}+(\chi_H-\breve{\chi}_H)\hat{H}$. For expositional ease we focus in the remainder on the endogenous factors in our model, hence $\hat{L}=\hat{H}=0$ (more details are provided in Appendix B). Because factor income shares are used in growth accounting to approximate the χ's it is impossible to distinguish $\breve{\chi}_H$ from χ_H, as our premise is that u is unobservable. The difference between input-adjusted and measured productivity growth is then:

$$\hat{A} = \hat{D} - \breve{\chi}_H \hat{u} . \qquad (4.24)$$

Only in the case that $\hat{u}$ is zero, productivity *growth* is measured accurately. However, if $\hat{u}$ is negative after the technology shock, as skilled workers increase learning time, total factor productivity is underestimated if measured in the naive way. A computer revolution induces increased learning time which leads to an underestimation of productivity growth. As u is unobserved, the estimation bias has to be inferred indirectly, an issue to which we turn in the next section.

5.2 Playing and learning

A back of the envelope calculation for the 1980s reveals that the magnitude of the measurement error could be substantial. Think about potential computer users as white-collar workers or non-production workers, two concepts empirically sufficiently similar for our purpose. Also blue-collar workers use computers and hence spend time learning. We ignore blue-collar workers to get a conservative estimate of the measurement error. The 1979–89 average wage-bill share of white-collar workers is roughly 0.46 (see Berman, Bound and Griliches, 1994). Combining this with the total share of income for labour of 2/3, $\chi_H\,(\approx\tilde{\chi}_H)$, is approximately 0.31.

One can think of u as time spent on on the job learning to master computer technology. Think of all the 'useless' surfing on the Internet. 'Useless' should be read in terms of gathering information that is instantaneously valueless. But this surfing contributes to the stock of knowledge, that is, how to get information; an asset that pays off in the future. A similar point can be made using e-mail, learning to use a word processor, changing a complete department to use CAD (Computer Aided Design), and so on. The time used for learning on the job is more speculative: suppose people starting to use the computer begin to experiment. Suppose that this takes an hour on a working day. This number seems excessive for some of the examples above but it is a lower bound for the ones where optimal computer use requires a reorganization. If *all* white-collar workers started experimenting in the same year, $\hat{u}$ would be -15 per cent, and productivity growth would be underestimated by 4.65% (0.31*15 per cent).

More plausible is the assumption that not all white-collar workers start to learn at the same time. Using data on computer use by Krueger (1993), we can infer that 5 per cent of the white-collar workers in 1979 used a computer, whereas this number increased to 48.4 per cent in 1989.[19] This implies that on average 4.3 per cent of the white-collar workers start using the computer every year. Consequently, underestimation of annual productivity growth of 0.20 percentage points would result (0.043*4.65 per cent).[20]

Total factor productivity growth in manufacturing is on average 1.2 per cent (1979–92). Before the oil crisis this was 1.8 per cent on average, see Gullickson (1995). Taking only white-collar workers into account, our measure explains 33 per cent of the productivity slowdown of 0.6 percentag points. This back of the envelope calculation reveals that the impact of computers – decreasing productive time – could be substantial. Hence, it is not surprising that we do not observe increases in TFP yet.

The calculation above, though ad hoc and uncertain, reveals that the *input* measurement error can be substantial. The next section deals with the other measurement errors in relation to a computer revolution and provides a more structured measurement strategy.

5.3 The knowledge-output problem: A dual approach to growth accounting

An appropriate assessment of the viability of the whole economy requires a more comprehensive output concept, broad output, that includes the value of produced knowledge capital ($q_f \dot{F}$ in the model). *Output*-adjusted productivity growth measures the productivity of the entire economy, including the productivity of knowledge production within firms. It is easy to see that *output* adjusted or ideal TFP ($\hat{A}'$) exceeds *input* adjusted TFP, as there is a positive jump in the wage of the skilled workers (Appendix B provides a derivation):

$$\hat{A}' = \hat{A} + (1-u)\,\chi_H\,\hat{w}_H \,. \qquad (4.25)$$

Hence the input- and output-correction method are not identical. The reason is that skilled workers start to learn on-the-job more intensively because the knowledge-investment activity is more attractive. Hence, ignoring the knowledge production part of the economy in growth accounting implies that the part where productivity gains arise is excluded.

Putting all pieces together, (4.25) and (4.23) substituted in (4.2) yields:

$$\hat{D} - \hat{A}' = \breve{\chi}_H\,\hat{u} - (\chi_H - \breve{\chi}_H)\hat{H} - (1-u)\,\chi_H\,\hat{w}_H \,. \qquad (4.26)$$

Hence $\hat{D}$ underestimates the true productivity growth of broad output ($\hat{A}'$) for three reasons (remember $\hat{u}<0$ and $\chi_H > \breve{\chi}_H$). The first term on the r.h.s. is the input exaggeration; increased time doing research leads to an underestimation. The second term on the r.h.s. of (4.26) indicates the productivity effect related to the scale effect prevailing in most growth models.[21] The third term indicates productivity growth in the research sector. Substitute $\hat{w}_H = \hat{q}_f + \hat{Q} + \hat{f}$ therein to obtain additional insight: (i) a GPT increases research productivity and the shadow price of knowledge capital in terms of consumption goods increases (ii) productivity of knowledge production goes up as organizational or firm-specific capital ($\hat{f}$) increases. Thus, the Solow residual might record a slowdown while the economy actually is booming.

How about our premise that u is unobservable? Both the input and the output adjusted TFP measures that correct the naive TFP measure, equation (4.24) and (4.25), depend on u. However, the *output* adjusted or ideal TFP growth measure is easily approximated by applying the dual approach to growth accounting. The dual approach starts from either a cost function or the income-expenditure identity. The cost function exploits the logic that factor rewards (= factor costs) can only increase if productivity increases. Alternatively, as Appendix B shows, recognizing that for *all* that is produced production factors are paid yields the following relation for dual growth accounting:

$$\hat{A}^{I} \approx \chi_H \hat{w}_H + \chi_L \hat{w}_L + \chi_K \hat{r} \; . \tag{4.27}$$

The dual approach to growth accounting gives an indication of the viability of *all* production possibilities including firm-specific knowledge capital. Empirical research estimating TFP by the dual approach is not widely available.[22]

The exposition yields three lessons. First, the knowledge-*input* problem affects growth accounting if u is non-stationary. Second, dual growth accounting is suitable if the production of valuable unobservables (knowledge for instance) is important. Third, the upsurge in wages of skilled workers preludes higher future growth rates. That is, as skilled workers' wages increase and productivity growth is sluggish, either capital rental rates and unskilled wages should be declining or productivity measurement is limited to a concept that is overly narrow. We claim the latter is the case.

6. CONCLUSION

This chapter has argued that increased wage inequality could well be the consequence of the computer revolution. We constructed a very simple endogenous growth model in which skilled workers received a premium as a result of higher productivity in learning or R&D in order to cope with the new technology. Output growth falls on the impact of a computer revolution as skilled workers reallocate their time from producing to learning. Unskilled workers will experience an initial decrease in wage(s) (growth) because there are less complementary skilled workers producing. As the new technology is implemented, skilled workers gradually reshuffle their time and spend more and more time producing again. Consequently, wage inequality decreases as skilled workers cooperate with unskilled workers in production. Output growth will recover in the long run when the value of doing research decreases relative to producing as the new technology is implemented and the stock of productive knowledge increases. In the long run, the growth rate of the economy is permanently higher because time devoted to the assimilation of new technologies has become more productive. Wage inequality will be at the same level as before the computer revolution. We have shown that if one does not take into account that the computer revolution has decreased production time, one seriously underestimates the rate of technological change. Here, we have given a clue to why wage inequality has increased due to the computer revolution and the fact that the computer revolution cannot be traced back in the productivity statistics.

NOTES

1. A decrease in the quality of those who leave the educational system early could, however, explain increased inequality. Chapter 6 provides a mechanism where increased supply of skill *is* increasing wage inequality.
2. This is controversial. DiNardo and Pischke (1997) and Oosterbeek (1997) argue that the relation is spurious. However, Autor, Katz and Krueger (1998) reconfirm earlier findings of Krueger (1993).

3. Wood (1998) argues that in the 1980s an *acceleration* in the growth rate of relative demand for high-skilled workers has occurred which is explained by trade with low-wage countries. Wood finds that biased technological change caused a *constant* increase in relative demand for skilled workers ever since the 1940s: relative supply of skilled workers grows steadily at 3 per cent a year. In a postscript to the paper he reports that the annual growth rate of relative demand for skilled workers was 3.8 per cent in the 1970s, 4.8 per cent in the 1980s and 3.0 per cent in the 1990s.
4. Obviously though, if high- and low-wage countries are no longer within the cone of diversification, increased trade with low-wage countries no longer affects inequality.
5. The acceleration in skill-biased technological change, loosely defined as both the factor and sector bias, does not necessarily need to be accompanied by an increase in TFP growth. For an extensive discussion of this point see Krugman (1995b). However, an acceleration in skill-biased technological change with a slump in TFP growth is often thought to be implausible.
6. This chapter reconfirms analytically the numerical results presented in Chapter 3.
7. Some of these issues are also discussed in Appendix A to Chapter 12 of Aghion and Howitt (1998a). The knowledge *output* problem is referred to as the knowledge investment problem by Aghion and Howitt.
8. For later reference we assume $\breve{Y}=Y$, as observed tangible output equals actual tangible output by assumption.
9. See Appendix E in the previous chapter for a set-up to analyse these issues. Howitt (1998) discusses measurement problems with respect to knowledge investment and focuses on endogenous depreciation rates. We differ from Howitt – apart from the modelling approach in the next section – in relating the knowledge investment problem explicitly to dual and primal growth accounting and relating the first to the second moment of the wage distribution.
10. Time subscripts are suppressed where it leads to no confusion.
11. The model would become analytically intractable if one allowed for capital as an input in the technology production function. However, the qualitative results that derive from the model with physical inputs would not be altered substantially compared to the model without physical inputs. See for example Mulligan and Sala-i-Martin (1993) for an exposition of the Lucas model with and without the inclusion of physical capital in the production function of human capital.
12. We abstract from an intertemporal utility maximizing framework here to reduce the number of differential equations and, thereby, the complexity of the model. See Chapter 3 for a model with Ramsey consumers.
13. Recall N is set to unity.
14. This condition is a modified version of the transversality condition known from the Ramsey model that the saving rate must be less than the share of capital or: $\alpha > s$, see e.g. Barro and Sala-i-Martin (1995, p. 89).

15. Alternatively, the stability condition could be rewritten as $(1-u)(1-\Theta)(1-sp)>(1-\alpha)$, which is derived from the transversality condition stating that f cannot grow faster than the interest rate r in the steady state.
16. The same reasoning holds for all kinds of R&D activities and for costly, technological-opportunity driven, adjustments. These costs are likely to cause an over-estimation of labour (research time) and capital (investments that are adjustment costs) inputs in national accounts.
17. Barro (1999) ignores the knowledge-input problem by focusing on the productivity of the gross output stream (that is, gross of intermediates and R&D expenditures). His simplifying assumption that physical output is the input of the knowledge production function serves analytical convenience but hides a serious measurement problem.
18. Again χ is used to denote income shares as we assume that production function F is similar to G.
19. David (1991) indicates 1977 as the year of the introduction of the microcomputer, so in 1979 hardly anybody worked with a computer. Using 5 per cent of the people already acquainted with computers is probably reasonable.
20. There are actually many more subtleties involved in estimating this slowdown. For example the length of the learning period should be taken into account as well as whether the adoption of computers accelerates or decelerates in the period, because people who stop experimenting cause productivity growth to be *over*estimated again.
21. A different way of looking at (4.26), is by inserting the steady-state conditions $\hat{u}=\hat{H}=\hat{Q}=\hat{q}_f=0$ and noting that the productivity growth of broad output ($\hat{A}'$) is simply a weighted sum of productivity growth of narrow output $\hat{D}$ and productivity growth in research and learning, $\hat{F}$.
22. See Hsieh (2002) for an extensive discussion of the dual approach to growth accounting in the traditional applications. Here we stress the suitability of the dual approach in case unobservable output (knowledge) is produced.

APPENDICES

A. Solution of the model

A.1 Firms

The current-value Hamiltonian of the optimal control problem for firm j reads as:

$$\mathcal{H}_j = p_j x_j - w_H H_j - w_L L_j - rK_j + q_{f_j} Q(i) f_j (1-u_j) H_j \ . \tag{A.1}$$

Some additional notation simplifies the exposition: $\Theta \equiv \beta(1-\alpha)$ and $\Xi \equiv (1-\beta)(1-\alpha)$. The first- order conditions for an optimum are:

$$\frac{\partial \mathcal{H}_j}{\partial L_j} = p_j \left(\frac{\varepsilon - 1}{\varepsilon} \right) \Xi T K_j^{\alpha} f_j^{1-\alpha} (u_j H_j)^{\Theta} L_j^{\Xi - 1} - w_L = 0 \ , \tag{A.2}$$

$$\frac{\partial \mathcal{H}_j}{\partial K_j} = r \leftrightarrow p_j \left(\frac{\varepsilon - 1}{\varepsilon} \right) \alpha T K_j^{\alpha - 1} f_j^{1-\alpha} (u_j H_j)^{\Theta} L_j^{\Xi} = r, \tag{A.3}$$

$$\frac{\partial \mathcal{H}_j}{\partial u_j} = 0 \leftrightarrow p_j \left(\frac{\varepsilon - 1}{\varepsilon} \right) \Theta T K_j^{\alpha} f_j^{1-\alpha} (u_j H_j)^{\Theta} u_j^{-1} L_j^{\Xi} - q_{fj} Q(i) f_j H_j = 0 \ , \tag{A.4}$$

$$\frac{\partial \mathcal{H}_j}{\partial H_j} = p_j \left(\frac{\varepsilon - 1}{\varepsilon} \right) \Theta T K_j^{\alpha} f_j^{1-\alpha} (u_j H_j)^{\Theta} H_j^{-1} L_j^{\Xi} - w_H - q_{fj} Q(i) f_j (1-u_j) = 0 \ , \tag{A.5}$$

$$\frac{\partial \mathcal{H}_j}{\partial f_j} = rq_{fj} - \dot{q}_{fj} \leftrightarrow p_j\left(\frac{\varepsilon-1}{\varepsilon}\right)(1-\alpha)TK_j^{\alpha}f_j^{-\alpha}(u_jH_j)^{\Theta}L_j^{\Xi} + q_{fj}Q(i)(1-u_j)H_j = rq_{fj} - \dot{q}_{fj} \ . \tag{A.6}$$

In addition we have to impose the transversality condition:

$$\lim_{t\to\infty} f_j \exp\left[-\int_0^t r(v)dv\right] = 0 \ . \tag{A.7}$$

The second and third FOCs give the no-arbitrage condition for the allocation of time of high-skilled workers in the production of goods and learning:

$$\begin{aligned} w_H &= p_j\left(\frac{\varepsilon-1}{\varepsilon}\right)\Theta TK_j^{\alpha}f_j^{1-\alpha}(u_jH_j)^{\Theta-1}L_j^{\Xi} \\ &= q_{fj}Q(i)f_j \ . \end{aligned} \tag{A.8}$$

A.2 Equilibrium

First, the differential equation for $R \equiv F/K$ can be obtained using the economy's resource constraint:

$$\frac{\dot{R}}{R} = Q(i)(1-u)H - spTR^{1-\alpha}(uH)^{\Theta}L^{\Xi} \ . \tag{A.9}$$

The derivation of the differential equation describing u requires two additional steps. First, the no-arbitrage condition (A.5) can be differentiated with respect to time to arrive at:

$$\alpha\frac{\dot{R}}{R} + (1-\Theta)\frac{\dot{u}}{u} = -\frac{\dot{q}_f}{q_f} \ . \tag{A.10}$$

Second, we can substitute the first term in FOC (A.6) out by rewriting the no-arbitrage condition:

$$\frac{1}{q_f}(1-\alpha)TR^{-\alpha}(uH)^{\Theta}L^{\Xi}=\frac{1-\alpha}{\Theta}Q(i)uH. \tag{A.11}$$

Using the last three results we obtain the differential equation in u:

$$\frac{\dot{u}}{u}=\left(\frac{1-\alpha}{\Theta}\right)Q(i)uH+\left(\frac{1-\alpha}{1-\Theta}\right)Q(i)H+\left(\frac{\alpha(1-sp)}{\Theta-1}\right)TR^{1-\alpha}(uH)^{\Theta}L^{\Xi}. \tag{A.12}$$

A.3 Stability

The stability of the equilibrium can be checked by evaluating the determinant of Jacobian matrix J at the equilibrium E:

$$J=\begin{bmatrix} \frac{\partial\dot{R}}{\partial R} & \frac{\partial\dot{R}}{\partial u} \\ \frac{\partial\dot{u}}{\partial R} & \frac{\partial\dot{u}}{\partial u} \end{bmatrix}. \tag{A.13}$$

The four partial derivatives of J at E are:

$$\left[\frac{\partial\dot{R}}{\partial R}\right]_E=-(1-\alpha)Q(i)(1-u^*)H<0\ , \tag{A.14}$$

$$\left[\frac{\partial\dot{u}}{\partial u}\right]_E=2\left(\frac{1-\alpha}{\Theta}\right)Q(i)Hu^*+\left(\frac{1-\alpha}{1-\Theta}\right)Q(i)H-(\Theta+1)\Phi><0\ , \tag{A.15}$$

$$\left[\frac{\partial \dot{R}}{\partial u}\right]_E = -Q(i)HR^*\left(1+\Theta\frac{1-u^*}{u^*}\right)<0 \ , \tag{A.16}$$

$$\left[\frac{\partial \dot{u}}{\partial R}\right]_E = -(1-sp)\left(\frac{1-\alpha}{1-\Theta}\right)\frac{r^* u^*}{R}<0 \ , \tag{A.17}$$

where $\Phi \equiv \frac{\alpha(1-sp)}{1-\Theta}TR^{*1-\alpha}(u^*H)^{\Theta}L^{\Xi}$. From the $\dot{u}=0$ locus follows that at E we have:

$$\Phi = \frac{1-\alpha}{\Theta}Q(i)Hu^* + \frac{1-\alpha}{1-\Theta}Q(i)H \ . \tag{A.18}$$

The equilibrium is saddle-point stable if $\partial\dot{u}/\partial u>0$; then the determinant of the Jacobian is negative. This will be the case if:

$$\left(\frac{1-\alpha}{\Theta}\right)Q(i)Hu^* - \frac{\Theta}{u^*}\Phi>0 \ . \tag{A.19}$$

Substitution of Φ gives:

$$u^*>\left(\frac{\Theta}{1-\Theta}\right)^2 . \tag{A.20}$$

B. Accounting

B.1 Accounting

The following assumptions with respect to the accounting procedure are necessary to obtain consistency. We assume the following: (1) Households own the shares of firms producing differentiated goods. (2) The manager of the firm decides on investment in firm-specific capital after which (3) firms pay profits ($\pi_i = x_i p_i - rK_i - w_L L_i - w_H H_i$) to households. (4) Firm i pays rental rate r to the capital assembly firms who pay this as dividends to the household.(5) Households decide on saving over total income (wage income, dividend and profits).

Firm *i*

$r\,K_i$	$p_i\,C_i$
$w_L\,L_i$	$p_i\,I_i$
$u\,w_H\,H_i$	
$(1\text{-}u)w_H\,H_i$	
π_i	
$S_f\,(\equiv q_f\,dF/dt)$	$(1\text{-}u)w_H\,H_i$

Households

$p\,C$	D
S	$w_H\,H$
	$w_L\,L$
	π

K-firms

D	$r\,K$
$P_I\,I$	S
dK/dt	$P_I\,I$

Note that due to these assumptions total investment (below the dashed lines) is equal to savings of households and savings generated internally in the firm. The assumptions with respect to accounting (and hence the model) could, without altering the main results, be changed such that total investment equals total household saving. This would, however, unnecessarily complicate the analysis of the dynamics of the model.

B.2 Standard growth accounting

Standard growth accounting ignores the accumulation of firm-specific knowledge capital and hence does not take into account investment time for firm-specific knowledge capital. So start with a standard production function:

$$\breve{Y}=DG(K,L,H) \; , \tag{B.1}$$

where K , L, and H measure physical capital, unskilled and skilled labour respectively. The growth of output is associated with growth in production factors and technology. By differentiating with respect to time, dividing by $\breve{Y}$ (=Y) and rewriting this yields:

$$\hat{D}=\hat{Y}-\chi_K\hat{K}-\chi_L\hat{L}-\chi_H\hat{H} \; , \tag{B.2}$$

where $\chi_i \equiv (D(\partial G/\partial i)i)/Y$, $i=K,L,H$. In practice the χ's are approximated by the factor income share, which is appropriate if the social marginal product equals the factor prices. The model presented in the main text does not have any externalities related to factor inputs. Therefore, the approximation is indeed appropriate.

B.3 Growth accounting with endogenous growth and firm-specific capital

The accounting procedure presented in Appendix B.1 shows that alongside the physical output, Y $(Y=C+I)$, knowledge capital is produced. Define the sum of these as broad output: Y^B. Obviously total production equals total income, see also Appendix B.1 (note that profits are ignored in the remainder), therefore:

$$Y^B = w_L L + w_H H + rK \ . \tag{B.3}$$

Differentiating the definition for income-production equality yields, after rearranging:

$$\dot{Y}^B = w_L L(\hat{L} + \hat{w}_L) + w_H H(\hat{H} + \hat{w}_H) + rK(\hat{r} + \hat{K}) \ . \tag{B.4}$$

Noting that total production includes knowledge production ($Y^B = Y + q_f \dot{F}$), differentiating this equation with respect to time yields:

$$\dot{Y}^B = \dot{Y} + \dot{q}_f \dot{F} + q_f \ddot{F} \ . \tag{B.5}$$

Differentiating the accounting identity, $q_f \dot{F} = (1-u) w_H H$, and substituting this into equation (B.5) and substituting the result into (B.4) and dividing by Y yields:

$$\begin{aligned} &\hat{Y} + \frac{1}{Y}\Big((1-u)(H\dot{w}_H + w_H \dot{H}) - \dot{u} w_H H\Big) = \\ &\frac{w_L L}{Y}(\hat{L} + \hat{w}_L) + \frac{w_H H}{Y}(\hat{H} + \hat{w}_H) + \frac{rK}{Y}(\hat{r} + \hat{K}) \ . \end{aligned} \tag{B.6}$$

Rewriting yields:

$$\begin{aligned} &\hat{Y} + (1-u)\frac{w_H H}{Y}\left(\hat{w}_H + \hat{H} - \frac{u}{1-u}\hat{u}\right) = \\ &\frac{w_L L}{Y}(\hat{L} + \hat{w}_L) + \frac{w_H H}{Y}(\hat{H} + \hat{w}_H) + \frac{rK}{Y}(\hat{r} + \hat{K}) \ . \end{aligned} \tag{B.7}$$

Now introduce the assumption that the χ's equal the income shares and substitute (B.2) into the previous expression and rewrite to get:

$$\frac{w_L L}{Y}\hat{w}_L+\frac{w_H H}{Y}\hat{w}_H+\frac{rK}{Y}\hat{r}=\hat{D}+(1-u)\frac{w_H H}{Y}\left(\hat{w}_H+\hat{H}-\frac{u}{1-u}\hat{u}\right)\ , \qquad \text{(B.8)}$$

which is the equation following from the equality of (4.25) and (4.27) in the main text. The expression on the r.h.s. is closely related to the perfect measure for productivity growth. The perfect measure for productivity growth is derived directly from the macroeconomic production function used in the model:

$$Y^B=AF(K,L,uH)+q_f Q(i)F(1-u)H\ . \qquad \text{(B.9)}$$

In growth rates:

$$\begin{aligned}\hat{Y}^B=\ &\frac{AF(K,L,uH)}{Y^B}\hat{A}+\psi_K\hat{K}+\psi_L\hat{L}+\psi_{uH}(\hat{u}+\hat{H})\\&+\frac{q_f Q(i)(1-u)H}{Y^B}\left(\hat{q}_f+\hat{Q}(i)+\hat{F}+\hat{H}-\frac{u}{1-u}\hat{u}\right)\ ,\end{aligned} \qquad \text{(B.10)}$$

where $\psi_i\equiv AF_i i/Y^B$, $i=K,L,uH$. Use $w_h=q_f Q(i)F$ (see (A.8)) to get $\hat{w}_H=\hat{q}_f+\hat{Q}+\hat{F}$ and substitute this in the previous expression. Bringing all factors relating to changes in factor inputs to the l.h.s. yields an expression for productivity growth.

$$\begin{aligned}&\hat{Y}^B-\psi_K\hat{K}-\psi_L\hat{L}-\psi_{uH}\left(1+\frac{1-u}{u}\hat{H}\right)\\&=\frac{AF(K,L,uH)}{Y^B}\hat{A}+\left(\psi_{uH}-\frac{uw_H H}{Y^B}\right)\hat{u}+\psi_{uH}\frac{w_H}{AF_{uH}}\frac{u}{1-u}\hat{w}_H\ .\end{aligned} \qquad \text{(B.11)}$$

Now substitute the factor prices for the marginal products. This gives us the following expression for the perfect measure for productivity growth of broad output:

$$\hat{A}' \equiv \hat{Y}^B - \frac{rK}{Y^B}\hat{K} - \frac{w_L L}{Y^B}\hat{L} - \frac{w_H H}{Y^B}\hat{H} = \frac{AF(K,L,uH)}{Y^B}\hat{A} + \frac{(1-u)w_H H}{Y^B}\hat{w}_H \,. \tag{B.12}$$

This expression for true productivity growth ($\hat{A}'$) has an obvious interpretation: productivity growth of broad output is the weighted sum of the increase in the technology of physical production and the increase in productivity of knowledge production. The latter is in turn equal to the change in the productivity of knowledge (as $\hat{w}_H = \hat{q}_f + \hat{Q} + \hat{f}$): that is the increase in organzsational or firm-specific capital ($\hat{f}$), the productivity effect of the GPT, plus the change in the shadow price, which is the value of knowledge capital in consumption goods. The weights are the factor-income shares in broad output. Our final task is to relate $\hat{A}$ to our observable $\hat{D}$. Use $\breve{Y} = DG(K,L,H)$, $Y = AF(K,L,uH)$ and $\breve{Y} = Y$ to arrive at:

$$\hat{D} = \hat{A} + \breve{\chi}_H \hat{u} - (\chi_H - \breve{\chi}_H)\hat{H} \,, \tag{B.13}$$

where $\breve{\chi}_H \equiv (AF_{uH}uH)/Y$. This expression prevails in the main text as the knowledge-investment-*input*-corrected growth account. Multiplying (B.12) by Y^B/Y and substituting (B.13) for $\hat{A}$ yields:

$$\frac{Y^B}{Y}\left(\hat{Y}^B - \frac{rK}{Y^B}\hat{K} - \frac{w_L L}{Y^B}\hat{L} - \frac{w_H H}{Y^B}\hat{H}\right) = \hat{D} + \frac{(1-u)w_H H}{Y^B}\hat{w}_H + \breve{\chi}_H \hat{u} - (\chi_H - \breve{\chi}_H)\hat{H} \,. \tag{B.14}$$

Substituting in (B.14)'s χ's that $AF_{qH} = w_H$ (which is easily verified to hold in the model) yields on the r.h.s. exactly the expression derived by the dual growth account (r.h.s. of (B.8)).

$$\frac{w_L L}{Y}\hat{w}_L+\frac{w_H H}{Y}\hat{w}_H+\frac{rK}{Y}\hat{r} =\hat{D}+(1-u)\frac{w_H H}{Y}\left(\hat{w}_H+\hat{H}-\frac{u}{1-u}\hat{u}\right)$$
$$=\frac{Y^B}{Y}\left(\hat{Y}^B-\frac{rK}{Y^B}\hat{K}-\frac{w_L L}{Y^B}\hat{L}-\frac{w_H H}{Y^B}\hat{H}\right) \quad \text{(B.15)}$$
$$=\frac{Y^B}{Y}\hat{A}'.$$

Hence the dual growth account overestimates the growth rate, as it is the ideal measure times Y^B/Y, and $Y^B>Y$. It is obvious that the dual growth account overestimates the growth rate as the weights on the l.h.s. of equation (B.8) are too high as their sum exceeds one (use $Y^B>Y$ and equation (B.3) to see this). However, as mentioned above, in practice the factor income shares that sum to unity are used, and hence the dual approach tends to the perfect measure.

5. Vested interests and resistance: Adopting a General Purpose Technology

1. INTRODUCTION

Do firms in the same industry apply the same and best-practice technology? Do all countries use the same technology? The answers to these questions are obvious: technologies employed by different firms and in different countries differ tremendously; even in modern times with the huge advances in travel and communication technology. 'Germans have been shocked by the weak technical standards in Britain' is not a quote from decades or centuries ago, but actually from 1998.[1] Such differences in technology are hard to understand by relying on the standard view of technology that treats it as a public good accessible to everybody at no cost. The question becomes urgent of why not everybody is employing best-practice technologies, even though these technologies are known and yield, by standard means of calculation, a positive net present value.

Several explanations have been explored. First, technology could be localized, meaning that the technology is available but useless with the given production technique.[2] This perspective explains why a seemingly better technology may actually not be superior in specific circumstances. Such a perspective might shed light on cross-country differences between widely different countries, but does not explain why in seemingly similar countries best-practice technologies are not implemented. Second, asymmetric information resulting in capital market imperfections, or information costs resulting in bounded rational behaviour might explain seemingly sub-optimal behaviour. Third, uncertainty with respect to future pay-offs may prevent firms from investing given that the investments are often to a large extent irreversible (for example, Dixit and Pindyck, 1994).

The perspective taken in this chapter is one in which we argue that adoption of a major new technology is costly. We refer to such a technology – that is useful to a wide variety of applications in the economy and requires adoption cost – as a GPT (Chapter 3 discusses the concept in greater detail). Jovanovic (1997) argues, for example, that adoption costs outweigh invention costs by roughly a factor of 20 or 30. Also, evidence that diffusion lags for new technologies are long suggests that sizeable adoption costs are at work. Opposite to common approaches (for example, Parente 1995, Parente and Prescott, 1994), we focus on the costs that workers have to bear. Alongside firms' installation and instruction costs, workers have to learn to use the new technology. An example is a training programme for a new technology that will be offered to employees during normal leisure hours instead of during working time. In the latter case, training hours effectively reduce production time so that firms bear the costs in terms of foregone production. In the former case it is the employee who has less leisure time left. Hence, the distribution of the adoption costs among different groups of workers, as well as among firms on the one hand and workers on the other is relevant. In this chapter, we focus on these distributional issues.

The story this chapter tells is simple. First, the net benefits from adopting a GPT may be unevenly distributed among workers when workers are heterogeneous. The heterogeneity is captured by differences with respect to age. Older workers could be more hesitant to a new technology since they have to go through a costly learning process, whereas they are not able to reap the full benefits from working with the new technology. Young workers are more willing to accept a new technology since they can recover the cost of training more easily, simply because their remaining lifetime is longer. Second, the share that workers get of the rent that is associated with improving the technology will generally differ from the share that the firm gets. This difference may depend on, for example, the competitiveness of the industry in which firms are operating (an idea that is also present in Aghion and Howitt, 1998b, and Prescott, 1998). Once a majority of the workers calculates that the net benefits of the new technology are negative, they will block the GPT by a majority vote over the allowance for the new technology. We thereby explicitly recognize that a firm is a collection of individuals, which cannot be modelled as an entity with a single mind. Although we model the decision to adopt a technology by voting, one might as well think of it as a decision influenced by resistance activities. In the most general interpretation, these activities include threats of worker disobedience, threats of consumer boycotts, loss of political sympathy, and so on.

The idea that new technologies or innovations are met with resistance is not new. As Machiavelli (1961, p. 51) pointed out, 'There is nothing more difficult to arrange, more doubtful of success, and more dangerous to carry through, than to initiate a new order of things... Men are generally incredulous, never really trusting new things unless they have tested them by experience'. In a similar spirit, Schumpeter argued that 'in the case of something new being attempted, the environment resists while it looks on with – at least – benevolent neutrality at repetition of familiar acts. Resistance may consist in simple disapproval – of machine-made products, for instance – in prevention – *prohibition* of the use of new machinery – aggression – smashing new machinery' (Schumpeter 1939, p. 100, italics added). Though in our formal analysis the focus is on voting over prohibition, the broader interpretation should be kept in mind when evaluating the results.

The basic conclusion we arrive at in this chapter is that GPTs of which the associated adoption costs are borne by workers are unlikely to be adopted. Moreover, in anti-competitive environments it is more likely that a new technology is resisted as the share of the rent associated with the new technology that accrues to workers is relatively small, making them more hesitant towards adoption.

The analysis adds to the literature on population ageing by emphasizing the relation between age distribution and technology adoption: economic growth may be reduced in an ageing society as the willingness to adopt new technologies is decreasing in remaining lifetimes. Hence, the chapter provides a complementary perspective on problems related to the age distribution of the population – besides issues in relation to savings and pensions.

The analysis in this chapter proceeds as follows. The next section discusses historical and anecdotal evidence on resistance to technological change. Section 3 presents the model we develop to analyse the adoption of technology and discusses the most important assumptions. The basic framework is as simple as possible. Hence initially we abstain from issues related to entry and exit of firms and endogenous costs of learning to use the new technology. These assumptions will be discussed and relaxed later on. The trade-offs different generations face and the way in which the generational composition of the economy affects the likeliness of a technology being blocked by workers is discussed in Section 4. Attention will also be paid to endogenizing learning costs. Section 5 evaluates the impact of competition policy. Having discussed the basic model, Section 6 analyses the model under free entry (and thus an endogenously determined number of firms). Section 7 concludes and discusses some potential extensions of the simple framework developed in this chapter.

2. RESISTANCE TO TECHNOLOGY

This section argues, by means of anecdotal and descriptive historical evidence, that resistance to technology is a potentially powerful explanation for the lack of implementation of 'best-practice' technologies. A convenient framework for the analysis of resistance to technological change distinguishes three separate but related layers. First, as technological progress can be viewed as a positive-sum game, it is not immediately obvious that resistance should arise at all. For resistance to arise there should be losers alongside the winners.[3] Hence, the first layer explores the underlying incentives for resistance activities. The second layer explores the action of resistance, hence whether the incentive for resistance fulminates in, for example, violence, regulations or unionism. The third layer consists of the consequences of the actions taken, namely the dynamic and static inefficiencies that arise due to effective resistance.

Section 2.1 explores, by means of analysing descriptive (historical) material, mostly the second perspective, that is, what type of resistance will lead to non-adoption of technology. Section 2.2 discusses theoretical approaches to the economics of technology resistance. This is a literature that deals rather implicitly with the actual resistance activity and focuses on the incentives and resulting inefficiencies.

2.1 Resistance

The failure to adopt new technologies may derive from various sources. The first, and historically very relevant, obstacle to the adoption of new technology is violence and effort to get political acclaim for the request to stick to the status quo in production technology. A difference in the extent and success of resistance to technology is one of the, admittedly many, pieces of the puzzle of why the Industrial Revolution occurred in Britain and not on the Continent. A piece, moreover, that is 'barely... explored' and 'dimly understood' (Mokyr, 1990, p. 255, 266). The Industrial Revolution brought gains to consumers and a few innovators, while the costs were mostly borne by skilled artisans. On the Continent 'the old urban guilds became a fetter on technological progress' (p. 258) and workers rioted against spinning machines imported from Britain.[4] The guilds resisted less by outright violence and riots, but more through a vast body of regulations and restrictions. Though resistance was eventually blown over, it retarded the introduction of new technologies on the Continent substantially. A potential explanation for less successful resistance in Britain is the fact that the ruling class had most of its assets in real estate and agriculture; assets whose value was not threatened by technological progress. Hence, the British ruling class had no interest in granting workers' claims for resistance (Mokyr, 1990).[5]

Second, obstacles to the adoption of new technologies may be unions that, by acting in favour of their members, resist labour-saving projects. Examples are found in the textile industry in India (Wolcott, 1994) where workers had less incentive to adapt and more ability resist to the introduction of labour saving technologies. Another example is the United States coal mining industry where strong unions were able to resist technological breakthroughs until they were pressed to give up their resistance due to the availability of cheap oil (Prescott, 1998).

A third, currently most pressing, source of the failures to adopt best-practice technologies derives from regulations and laws that formally prevent technological improvements. Their importance is well established in Baily (1993) and Baily and Gersbach (1995). Baily studies productivity differences in service industries[6] operating in Europe, the United States and Japan, and evaluates the role of certain types of regulation and the intensity of competition as a potential explanation for observed productivity differentials. The general lesson to be drawn is that incomplete convergence in a service industry's productivity is due to regulatory barriers and lack of competition, causing slack and inhibiting productivity growth. Such barriers could be created by special interest groups trying to protect their economic interests. A case study of banking stresses the idea that available innovations may not be adopted without the threat of competition. Competition in the banking industry is limited by the regulated structure of this sector, resulting from the fact that most countries considered financial stability more important than economic (competitive) efficiency. In the United States, however, the regulatory environment has encouraged competition. Overall productivity varies widely in the countries studied. Germany's overall productivity level in banking is about 68 per cent of that in the United States. This productivity gap is due to the failure to exploit economies of scale and scope in German banks with their many small branches, and their less effective use of information technology. Part of the productivity differential between German and United States banks is due to the fact that far fewer transactions per person pass through ATMs in Germany than in the United States, so that German banks need more labour input to dispense cash to their customers. The fact that new technologies such as ATMs were adopted earlier in the United States reflects that competition in the United States banking industry is more intense, forcing banks to cut operation costs and streamline their operations.

International productivity differentials in manufacturing are studied in Baily and Gersbach (1995). The major part of the observed productivity differences is caused by the way functions and tasks are organized, and the fact that some companies require less labour and material input by new designs of their products.[7] The question then arises of why companies fail to exploit best-practice technologies. An illustrative example is that German breweries are simply forbidden to adopt better technologies that are used in the United States and Japan. Baily and Gersbach suggest that exposure to the most efficient producers through global competition will speed up the adoption of new technology and promote efficiency.

Besides all mechanisms driven by some explicit loss, one might argue that 'In every society, there are stabilizing forces that protect the status quo. Some of these forces protect entrenched vested interests that might incur losses if innovations were introduced, others are simply don't-rock-the-boat kinds of forces. Technological creativity needs to overcome these forces' (Mokyr, 1990, p. 12). How strong and pervasive these pressures can be, exerted in the old days by guilds, has been nicely illustrated by Olson (1982) in his book on *The Rise and Decline of Nations*. In his description of the operation of the caste groups in India, he argues that these groups controlled lines of business, kept crafts mysteries or secrets, used boycotts and strikes and were organized with a strong emphasis on obedience to the rules of the caste ruling out competition.

Despite overwhelming evidence of the relevance of resistance activities, more formal analysis is scarce. A sketch of the available contributions will be discussed in the next section.

2.2 Related theoretical literature

As mentioned earlier, the theoretical literature on technology adoption mainly focuses on the adoption costs for firms. One important exception is Helpman and Rangel (1999). They focus on the short-run response to an arrival of a

GPT, and answer the central question of under what conditions the arrival of a GPT causes a decrease in aggregate productivity. At the heart of the analysis is a (continuous) overlapping-generations framework. Two basic effects are distinguished, namely (i) an adoption effect and (ii) an entry effect. The adoption effect results from the fact that the introduction of a new technology makes old experience worthless, implying a temporary drop in productivity for those who adopt. In addition there is an entry effect, which may be positive or negative. It is positive in the case that new technology requires less schooling and results in students leaving school earlier than expected (the case of technology-skill substitutability). The entry effect is negative in the case that new technology requires additional schooling and results in more people going to school instead of producing (the case of technology-skill complementarity). The general conclusion is that the arrival of a new GPT can trigger a temporary slowdown in growth. This is the more likely the higher the required educational level to operate the new technology, the more experience is lost upon adoption of the new technology and the faster experience is generated when using the new technology (making adoption more attractive).[8]

Adoption when technology arrives in a continuous stream, opposite to the GPT with a once-and-for-all nature in Helpman and Rangel (1999), is analysed by Chari and Hopenhayn (1991). In a two-generation OLG economy they show that vintage-specific human capital of the old, accumulated through learning-by-doing when unskilled and young, leads to slow adoption of new technology if newborn unskilled are highly complementary with human capital of the old.[9] Jovanovic and Nyarko (1996) show in a similar vein that in a one-agent model, even bounded learning-by-doing might be detrimental to growth. Abundance of expertise with a specific vintage that is non-transferable to a new vintage (specific human capital) locks the agent in.

Complementary to our work is Caselli (1999) who analyses adoption of technology in a model with heterogenous agents. The heterogeneity is reflected in the cost of learning to work with the next generation technology. Where we stress the heterogeneity on the benefit side (lifetime benefits depend on age), Caselli emphasizes the cost side by pointing at differences in ability.

A theoretical contribution that is closely related to ours is Holmes and Schmitz (1994, further H&S). They focus on the effects of trade on technology adoption. Workers are modelled as being skilled in that they have a technology-specific advantage in one sector of the economy. As this relative advantage is to be lost if adoption takes place, skilled workers might opt for resistance to a new technology. Blocking technology is modelled as a legal ban on consumption of goods produced with the newly available technology or on the use of the newly available technology. In a closed economy these two are identical. The cost of blocking a technology by skilled workers, by means of resistance, is weighted against the real income gain.[10] The main conclusion is that tradeability of goods acts as an important disciplinary device, as non-adoption, to protect specific rents, is ineffective because the rents will be competed away by foreign firms who will adopt (by assumption).

Our approach differs in several important ways from H&S. Firstly, we introduce heterogeneity by age whereas H&S distinguish workers who have or don't have technology-specific skills. This allows us to study the relation between age structure and technology adoption. Secondly, our approach differs from H&S with regard to competition. We have imperfect competition, whereas H&S have perfect competition. Furthermore, competition, in our perspective is (also) a domestic concept, whereas H&S only consider changes from autarky to free trade.

3. A SIMPLE FRAMEWORK

We study the adoption decision of a newly arriving GPT in a simple overlapping-generations model. Consider an economy with a large number of individuals. At time t, a generation is born of size P_t. At each point in time, there are three overlapping generations alive, the young, the middle-aged and the old. Total population size at time t is thus equal to $P_t+P_{t-1}+P_{t-2}$.[11] All generations work and are equally productive. Households are endowed with one unit of time, which they can use for working, leisure or education. Total leisure time (v) is thus equal to $1-l-s$, where l is labour time and s is schooling time. Consumers derive utility from the consumption of a homogeneous good, a composite of differentiated products and leisure. Besides the workers, the economy is inhabited with N holders of ideas.[12] The possession of an idea enables them to produce a differentiated good. They make profits which they spend on consumption (based on the same homothetic utility function that applies to all other consumers).

The present discounted utility (u_t) of an individual born at date t is given by:

$$u_t = \sum_{\tau=0}^{2} \delta^\tau \left[c_{t+\tau,t} + \nu \ v_{t+\tau,t}\right], \tag{5.1}$$

where $\delta<1$ is the discount factor and ν is the relative preference for leisure. Where two subscripts are added, the former denotes time whereas the latter denotes date born. Hence, $c_{t+\tau,t}$ is utility derived from consuming a bundle of goods at time $t+\tau$ by an individual born at time t and $v_{t+\tau,t}$ is leisure time consumed at time $t+\tau$ by an individual born at time t. We assume that consumption and leisure are perfect substitutes.[13] We assume that labour supply (l) is perfectly inelastic and equal to $\bar{l}$ (this assumption is relaxed in Canton, De Groot and Nahuis (1999)). The bundle of goods that is consumed consists of a numeraire good (y) and a composite of high-tech goods (labelled x)[14] which consists of N varieties of a quality differentiated high-tech good (which are indexed i=1,…,N). The instantaneous sub-utility function of an individual consumer with regard to the consumption good is specified in Cobb-Douglas format:

$$c = x^\beta y^{1-\beta} . \tag{5.2}$$

This homothetic utility function applies to individuals from all three generations alive who spend their wage income, and to the idea owners, who spend their profits. The composite x is defined as[15]

$$x = N^{\sigma}\left[\frac{1}{N}\sum_{i=1}^{N}\breve{x}_i^{\frac{\varepsilon-1}{\varepsilon}}\right]^{\frac{\varepsilon}{\varepsilon-1}}, \tag{5.3}$$

where $\breve{x}$ is the individually consumed amount of the high-tech good of variety i, σ represents the taste for variety, and ε measures the elasticity of substitution between any pair of differentiated high-tech goods. The numeraire good Y is produced according to a simple linear technology

$$Y = L_Y , \tag{5.4}$$

where L_Y is the amount of labour employed to produce the numeraire good. Labour productivity in this sector is, without loss of generality, taken to be equal to one. Each individual firm in the high-tech sector produces according to the following technology[16]

$$X_i = f_i L_{Xi}^{\mu} , \tag{5.5}$$

where X_i denotes the produced volume of the high-tech good of variety i, L_{Xi} is skilled labour employed in the i-th intermediate firm, f_i is its labour productivity, and μ indicates a scale effect. The labour market clears so

$$L_{Yt} + \sum_{i=1}^{N} L_{Xit} = L_t = \sum_{\tau=0}^{2} P_{t-\tau} l_{t,t-\tau} , \tag{5.6}$$

where L_t is total labour supply across all generations (in time units). Due to our normalization, we have L_t=$3P\bar{l}$=$\bar{l}$, which will further be denoted as L. Note that since workers from different age groups are perfect substitutes, we can simply add them up.

With a given number of idea holders the condition for labour market clearing completes the description of the model. Implicitly, a perfect entry barrier is present in the economy such that the number of firms/ideas is fixed. The sensitivity of our results for this assumption is checked in Section 6. In solving the model, however, we will proceed here under the assumption of blocked entry.

We can now straightforwardly solve the model for a given labour productivity and a given number of firms by solving (i) the two-step consumer optimization problem yielding the division of income over the goods available to the consumers and (ii) the profit-maximization problem for producers of the goods (see Appendix A.1 for a detailed description of the model solution). This results in an allocation of labour that looks like

$$L_{Xi} = \frac{\beta\mu(\varepsilon-1)L/N}{(1-\beta)\varepsilon+\beta\mu(\varepsilon-1)} \quad \text{and} \quad L_Y = \frac{(1-\beta)\varepsilon L}{(1-\beta)\varepsilon+\beta\mu(\varepsilon-1)}\,. \tag{5.7}$$

The size of firms in the high-tech sector will be larger, the larger the supply of labour (L) in time units is, the smaller the number of producers (N) is, the more weight consumers attach to the consumption of high-tech goods (β), the smaller the distortion due to mark-up pricing (that is, when ε is increased) is, and the more important economies of scale (μ) are.

4. THE DECISION WHETHER OR NOT TO ADOPT A NEWLY ARRIVED GPT

Having established the equilibrium solution of the model, we can now discuss how consumers will react to the arrival of a new GPT. Suppose that at time t_1 a new, better technology becomes available to produce the quality differentiated high-tech goods (f^A). When this technology is adopted, each worker has to incur a cost in terms of foregone leisure time equal to $s=\bar{s}$ at time $t=t_1$. We assume, for simplicity, that the technology can fully be employed in period t_1+1 and is γ times as effective as the old technology ($\gamma \equiv f^A/f > 1$). Upon effectively being used, the old technology is immediately and fully replaced by the new technology. Two remarks are in place. First, we make the assumption that the old incur the cost of adjustment even when they stand to lose from the adoption of the GPT. This assumption can be justified by the simple fact that change is disliked.[17] Second, we assume that income transfers between generations and between the holders of ideas and workers (other than wage payments) are impossible. This is due to a hold-up problem present in the model. Neither people from the current young generation nor the possessors of ideas can credibly commit to transfer money since they have no incentive to do so once the investment has been made (under the absence of perfect contractibility).[18]

Let us now characterize the costs and benefits of adoption for each age group.[19] We are allowed to ignore the generation that enters the labour market in the period after the adoption decision is taken for two reasons. First, we think of that generation as those currently enrolled in primary and secondary education, hence their education will be adjusted to the new dominant GPT. This means that there is no *additional* cost of adoption and change for them because they are well-prepared for the new GPT. Second, that generation has, in general, no suffrage and hence does not affect the outcomes of voting.

We denote the net change in utility that results from adoption at time t_1 for an individual borne at $t_1-\tau$ by $F_{t_1,t_1-\tau}$. So for the currently old, we get

$$F_{t_1,t_1-2} = \left[c^{A}_{t_1,t_1-2} - c^{NA}_{t_1,t_1-2}\right] - \nu\bar{s} \,, \tag{5.8}$$

where the indices *A* and *NA* refer to situations with and without adoption, respectively. In order to determine the value of this *F*-function, we need to establish the consumption index *c*. As shown in Appendix A.2, this value can be derived as

$$c = \bar{l}\left(\frac{\beta(\varepsilon-1)\mu f L_X^{\mu-1}}{\varepsilon N^{1-\sigma}}\right)^{\beta}(1-\beta)^{1-\beta} \,. \tag{5.9}$$

Since the new technology only becomes effective after one period and L_X is constant (see equation (5.7)), the consumption index is not affected by adoption and hence $F_{t_1,t_1-2}=-\nu\bar{s}$. So the currently old lose from adoption. Once the new GPT is implemented, they have passed away. Nevertheless, they have to incur the cost of foregone leisure. As a consequence, the currently old will be against the adoption of a new technology.

For the other two age groups, the returns from adoption are not necessarily negative. On the one hand they gain due to the fact that the GPT enlarges their (future) consumption basket. This follows from the fact that $\partial c/\partial f>0$ (see equation (5.9)). On the other hand, however, both groups lose as some leisure time has to be foregone. Without adoption an individual has $1-\bar{l}$ units of leisure, while with adoption individuals only enjoy $1-\bar{l}-\bar{s}$ time units of leisure. These costs are more likely to exceed the benefits for the middle-aged than for the young since the former can only benefit from the enlarged consumption possibilities for one period. This shows up in the F-functions for the young and medium aged, that respectively look like

$$F_{t_1,t_1} = [c^A_{t_1,t_1} - c^{NA}_{t_1,t_1}] - \nu\bar{s} + \delta[c^A_{t_1+1,t_1} - c^{NA}_{t_1+1,t_1}] + \delta^2[c^A_{t_1+2,t_1} - c^{NA}_{t_1+2,t_1}] , \qquad (5.10)$$

and

$$F_{t_1,t_1-1} = [c^A_{t_1,t_1-1} - c^{NA}_{t_1,t_1-1}] - \nu\bar{s} + \delta[c^A_{t_1+1,t_1-1} - c^{NA}_{t_1+1,t_1-1}] . \qquad (5.11)$$

From these two equations, it is evident that a sufficient condition for the young to be in favour of adoption is that the middle aged are in favour of adoption ($F_{t_1,t_1}>F_{t_1,t_1-1}$). Hence, under a system of majority voting the critical question that we need to address is whether the middle-aged are in favour or against adoption since the middle aged have the decisive vote. Using equation (5.9), we can derive that $c^A_{t_1}=c^{NA}_{t_1}$ and $c^A_{t_1+1}=\gamma^\beta c^{NA}_{t_1}$; recall that $\gamma\equiv f^A/f$. The middle aged are thus in favour of adoption provided that

$$F_{t_1,t_1-1} = \delta\left(\gamma^{\beta} - 1\right)c^{NA}_{t_1+1,t_1-1} - \nu\bar{s} > 0 \ . \qquad (5.12)$$

So adoption is preferable if the discounted benefit from accepting the GPT exceeds the associated cost in terms of foregone leisure.

We will now proceed by characterizing the increase in f that is required to make the group of middle-aged people just indifferent between adopting the new technology and proceeding with the old one. This critical increase will be labelled γ^*. Since $\partial F/\partial\gamma>0$ we know that there is exactly one value $\gamma^*>1$ for which middle-aged consumers are indifferent, *i.e.* $F=0$. Combining equations (5.9) and (5.12), the critical value of γ is easily derived as (see Appendix A.2)

$$\gamma^* = \left[1 + \frac{\nu\bar{s}\left(\varepsilon N^{1-\sigma}\right)^{\beta}}{\delta\bar{l}\left(\beta(\varepsilon-1)\mu f L_X^{\mu-1}\right)^{\beta}(1-\beta)^{1-\beta}}\right]^{1/\beta}, \qquad (5.13)$$

where L_X is determined by equation (5.7). Obviously, the required productivity gain to leave middle-aged workers indifferent is increasing in the time loss ($\bar{s}$). γ^* is decreasing in δ, as a loss today is more easily compensated tomorrow, if tomorrow is valuable.

The costs of adjustment (s) have so far been simply modelled as being independent of the nature of the GPT (γ). More realistically, we might model the cost of adoption as consisting of an exogenous part, and a part that is related to the size of the increase in productivity:

$$s = \bar{s} + a(\gamma^{b}-1) \ . \qquad (5.14)$$

The parameter $a>0$ reflects the idea that adoption becomes more costly if the increase in productivity becomes larger, while $b>1$ captures the idea that costs of adoption increase more than proportionally with the increase in productivity. The solution of the model can now conveniently be represented in a graphical manner as in Figure 5.1. The vertical axis in the graph depicts gross benefits of adopting the new technology for the middle-aged consumers (that is, the increase in utility associated with the higher productivity; the light dashed line), costs of adopting the new technology (the heavily dashed curve), and the net benefit (the solid line). The new technology will be adopted, provided that the net benefit is positive.

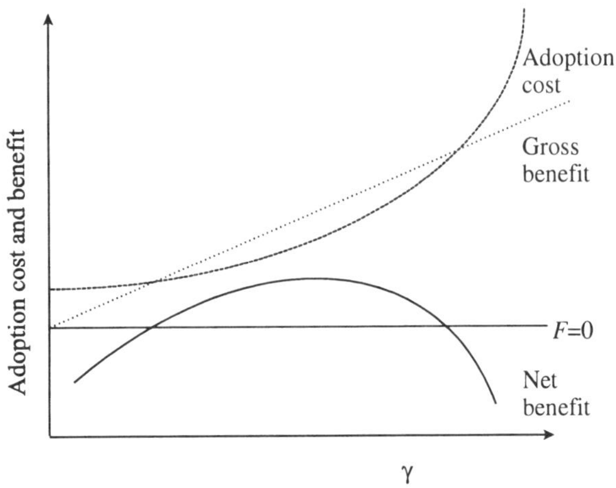

Figure 5.1 *Endogenous adoption costs*

The figure reveals that adoption will occur provided that (i) the increase in productivity is sufficiently large to enable the consumers to 'recover' their loss in utility associated with the costs of adoption, and (ii) the increase in productivity is not too large to prevent the adoption costs dominating the gross benefits. So improvements should not be too small, as change bears some fixed costs, but improvements should not be too large either as learning costs will become prohibitive. Any factor that will increase the gross benefits of adoption or decrease the costs will enlarge the range of technologies that will be adopted.[20]

5. THE EFFECTS OF COMPETITION ON TECHNOLOGY ADOPTION

In this section, we will discuss how a change in rivalry between firms producing the quality-differentiated product will affect workers' resistance to change. We will model enhanced competition as an increase in the substitution elasticity between high-tech goods. The question of whether competition will foster the adoption of new technologies and thereby foster economic growth then boils down to the question of whether the increase in productivity required to make middle-aged consumers equally well-off becomes smaller when competition increases. Hence, we have to determine whether $\partial\gamma^*/\partial\varepsilon$ is negative. From equation (5.13) we can straightforwardly show that competition will foster economic growth. This result is easily understood. An increase in productivity increases the total cake that can be divided between workers and the owners of ideas. In addition, the share of the increase in the cake that is obtained by the workers positively depends on the elasticity of substitution between high-tech goods (see Appendix A.3). If the elasticity of substitution is close to one, the owners of ideas receive almost all of the rents associated with the adoption of the new technology. As a consequence, resistance to adoption by workers will be large. If, on the other hand, most of the increase in the cake goes to the workers, resistance will be limited. This is also illustrated in Figure 5.2. An increase in competition shifts the gross benefit curve counter clockwise and hence lowers the critical productivity increase (see the dashed line).

6. FREE ENTRY AND THE EFFECTS OF COMPETITION ON RESISTANCE TO CHANGE

So far, we have studied the characteristics of the model under the assumption that the number of high-tech firms is exogenously given. This implies that persistent profits can exist; profits are not competed away by entry of new

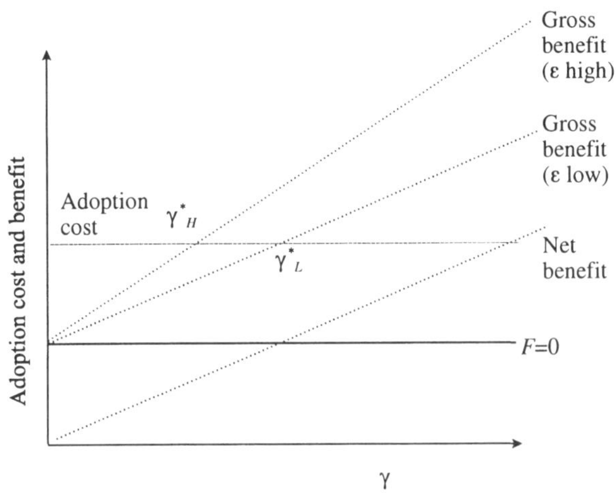

Figure 5.2 Enhanced competition

firms. In a model where knowledge is tied to firms, this assumption can be defended. However, it is interesting to consider the characteristics of the model once we drop this assumption. Such an exercise is interesting for various reasons. Firstly, it can be seen as a check for robustness of previously derived results. Secondly, it is interesting as it gives additional insights on the effects of competition on technology adoption.

The solution of the model under the assumption of free entry and exit is straightforward after introducing a fixed production cost[21] (for details see Appendix B). Following the same procedure as in Section 4, we can determine the critical increase in productivity that is required for middle-aged workers to be indifferent between adopting the new technology and sticking to the old one. This critical increase is again given by equation (5.13), where the allocation of workers to the high-tech sector and the number of firms are now given by

$$L_{Xi} = \frac{\mu(\varepsilon-1)L_f}{\varepsilon-\mu(\varepsilon-1)}, \tag{5.15}$$

$$N = \frac{\beta L}{L_f}\left[1 - \frac{\mu(\varepsilon-1)}{\varepsilon}\right], \tag{5.16}$$

where L_f is the fixed production cost. We restrict μ to be smaller than $\varepsilon/(\varepsilon\text{-}1)$ to guarantee an economically meaningful equilibrium with a positive number of firms and positive firm size. Equation (5.15) says that high-tech firms become larger, the larger the fixed cost and the stronger competition is. This is easily understood as large fixed costs and strong competition require firms to sell and produce a lot in order to break even. The number of high-tech producers is determined in equation (5.16). There are fewer firms active when fixed production costs are higher, and when economies of scale become more important.

By substituting the latter two equations into the solution for γ^* (equation (5.13)), we can obtain a closed-form expression for γ^* (see Appendix B). We are particularly interested in the effect of competition on this critical productivity shift. To that end, we need to evaluate

$$\frac{d\gamma^*}{d\varepsilon} = \frac{\partial\gamma^*}{\partial\varepsilon} + \frac{\partial\gamma^*}{\partial N}\frac{dN}{d\varepsilon} + \frac{\partial\gamma^*}{\partial L_{Xi}}\frac{dL_{Xi}}{d\varepsilon}. \tag{5.17}$$

The first term on the right hand side is the direct effect from competition on technology adoption, and the two other terms are the indirect effects. We discuss these effects in turn:

- The direct effect captures the same mechanism as described in Section 5: increased competition lowers the price of the high-tech good and thereby increases real income of workers which makes them more inclined to adopt the new technology. Since profits earned by the high-tech firms are driven down to zero, competition does not have an effect on the share of the cake that accrues to workers.
- The first indirect effect is only relevant in the free entry model as it operates through changes in the number of high-tech producers. We know

that $dN/d\varepsilon<0$. When competition is intensified, mark-ups decline and fewer firms will survive in the presence of fixed operation costs. Furthermore, $\partial\gamma^*/\partial N<0$; the reduction of product variety makes workers worse-off and hence the increase in productivity required to keep them equally well-off increases. This 'variety-effect' works against the direct effect.

- With respect to the second indirect effect, we know that $dL_{Xi}/d\varepsilon>0$; high-tech firms become larger when competition is intensified in order to recoup their fixed production costs. Furthermore, $\partial\gamma^*/\partial L_{Xi}<0$ if there are economies of scale in production of high-tech goods ($\mu>1$). So this 'scale-effect' has a similar sign to the direct effect. Competition increases firm size and results in better exploitation of economies of scale of the remaining firms, thereby reducing consumer prices and increasing their real income.

In contrast to Section 4, the relationship between growth and competition is no longer unambiguous. As shown in Appendix B, the relationship between γ^* and ε is only negative provided that the parameter measuring the taste for variety is smaller than the mark-up (that is, $\sigma<\varepsilon/(\varepsilon-1)$). In the special case in which the taste for variety and the mark-up are exactly equal, the elasticity of substitution does not affect the critical increase in productivity (this is the case that is explicitly considered by Dixit and Stiglitz, 1977). When the taste for variety is large, productivity should increase relatively strongly in more competitive environments to make middle-aged workers equally well off.[22] The intuition behind this result is as follows. An increase in competition makes high-tech firms larger and reduces the equilibrium number of high-tech firms (and thereby the variety of high-tech goods available to consumers). This makes consumers worse off. On the other hand, an increase in competition reduces the distortion due to mark-up pricing and increases real income of consumers, which makes them better off. When the taste for variety is low, the latter effect dominates the former and increased competition needs to be accompanied by a relatively low increase in productivity for the middle-aged worker to be equally well off. If, on the other hand, the taste for variety is strong, the former effect dominates and the increase in productivity required to make middle-aged workers equally well off upon adoption of the new technology is larger in more competitive environments.

7. CONCLUSIONS

Productive efficiency is largely determined by the applied technology. Two observations stand out: first, technologies actually implemented differ vastly across countries; second, within countries many technologies that would obviously improve firms' efficiency are not adopted. In this chapter, we provide an explanation for these observations. It is well documented that the adoption of a GPT often requires an investment by the firm. But the introduction of a new technology could also turn existing operation methods obsolete, and employees may have to learn how to operate the new technology. Since the latter effect is often ignored, we concentrated in this chapter on the benefits and costs of technology adoption for the worker who has to get acquainted with the new technology. If the costs of adoption for the worker exceed the benefits that can be reaped, workers will engage in activities aimed at keeping the old technology in place.

This chapter showed by anecdotal historical evidence that resistance to change might indeed be an important ingredient in the explanation of huge productivity differences across countries, as well as differences in technologies employed by different firms. Historical evidence shows that an important cause for blocked technology adoption is resistance of workers instead of resistance of firms.

We developed a simple overlapping-generations model to formalize these ideas. The model shows that the prime motive for resistance to technology comes from the fact that older workers have limited time to get a sufficiently substantial 'return on their investment'. Hence, ageing might also be a problem from this perspective. Moreover, it was shown that increased competition tends to increase the likeliness of adoption and hence accelerates growth, which is a well-documented phenomenon in descriptive material.

Apart from extensions discussed or mentioned earlier, there are several interesting ways in which the model could be extended and refined. The first is the introduction of imperfect substitution in production between labour from different age groups. A second extension would be to describe the bargain over the surplus associated with the adoption of the technology between the firm and its workers.

Although these extensions are interesting in their own right, they will not affect the basic mechanisms described in this chapter. What crucially matters for the question whether a GPT will be adopted is the size of the rent generated with the GPT, the speed with which this rent can be consumed and the way it is divided over various stakeholders within the economy.

NOTES

1. *The Economist,* 5 December 1998. About German BMW assessing productivity in the British Rover factories.
2. Localized technological progress (Atkinson and Stiglitz, 1969, Stiglitz, 1987, and Antonelli, 1995) or appropriate technology as Basu and Weil (1996) call it, provides a potential microeconomic foundation for the specificity of knowledge. The basic idea of localized learning is simple: learning-by-doing or innovation does not shift the *total* isoquant inward, but only the part that is actually used (hence where the initial technology choice was optimal given the factor prices). So with different factor endowments a poor country is unable to acquire total factor productivity levels of a rich country, as the technology requires the factor endowments of rich countries.
3. Though obvious, the necessity of distinguishing between groups that tend to lose and groups that tend to win from technological progress immediately reveals that a symmetric, homogeneous-agent framework is insufficient for our purpose.
4. Mokyr warns not to *identify* machine breaking, commonly observed early in the Industrial Revolution, with resistance to technology. More often this was simply a sign of a labour dispute, as machines were a vulnerable target.
5. Actually, in Britain destruction of machinery was made a felony punishable by death.
6. Four major industries are examined: airlines, retail banking, telecommunications, and general merchandise retailing.

7. Nine industries (automobiles, automotive parts, metalworking, steel, computers, consumer electronics, food, beer, soap, and detergent) in three countries (Germany, Japan, and the US) are studied. Baily and Gersbach find that physical capital and embodied technology can account for part of the productivity-differential. The major part of productivity differences, however, is to be found elsewhere (in technology and/or organization).
8. Some unattractive features of the model are that it is not a general-equilibrium model and that the old and the new technology can co-exist.
9. Krusell and Ríos-Rull (1996) introduce a dynamic political process in the Chari and Hopenhayn (1991) model and analyse stagnation and growth equilibria. Our one-shot voting approach simplifies the political process compared to Krusell and Ríos-Rull's dynamic voting process (where the current policies' consequences for future political equilibria are taken into account). We, however, extend their discussion by focusing on the general equilibrium effects of competition on regime switches from stagnation to growth.
10. Erecting a barrier increases the price of the manufacturing good relative to the no-barrier case, so that real income is decreased. However, skilled workers' income might increase if a barrier is erected. If skilled workers are relatively scarce, a barrier causes the price of the manufacturing good to be relatively high, as it is determined by the marginal rate of transformation of unskilled workers between the two goods (who are equally productive in both activities). In that case a skilled worker's marginal product is higher with a barrier. If skilled workers are relatively abundant, their income equals that of unskilled workers as they both produce the non-manufacturing good, in which skilled workers have no advantage.
11. In what follows, we assume, without loss of generality, that population growth is absent. This implies that $P_t = P$ at all points in time. In the remainder, we will normalize P at one-third for presentational convenience.
12. We abstain from a discussion of inheritance of patents or ideas. The number of idea owners is small compared to the total population and therefore irrelevant for the decision whether or not a new technology will be adopted (which is assumed to be the outcome of a vote among the consumers; see Section 4). However, the number of idea owners (and thus the number of varieties of the high-tech good produced in the economy) is assumed to be sufficiently large so that we can assume that firms compete monopolistically (for different modes of competition we refer to van de Klundert and Smulders, 1997). Moreover, for simplicity, we assume that idea owners do not supply labour on the labour market.
13. Imperfect substitution between consumption and leisure is discussed in the working paper version of this chapter, see Canton, de Groot and Nahuis (1999).
14. Where it leads to no confusion, we drop indices. Lower case variables pertain to individually consumed volumes (to be distinguished from macroeconomic volumes which are indicated with capitals).
15. A discussion of this aggregation function is beyond the scope of this chapter. See de Groot and Nahuis (1998) and Chapter 7.

16. All individual demands of the three generations alive and the idea owners sum up to total supply of the high-tech good (X_i).
17. In Chapters 3 and 4 a more structural approach to the temporary negative effects of implementation of a GPT is presented.
18. This implies that the adoption decision is in some cases not Pareto-efficient. Moreover, the economy is not efficient due to distortionary pricing in the monopolistically competitive sector.
19. Whether the possessors of ideas are in favour of adoption depends on two factors, namely (i) the effect on their nominal income (the profit π), and (ii) the purchasing power of the profit.
20. Another alternative for endogenizing the costs of learning would be the introduction of technology-specific knowledge. This technology-specific knowledge may be general (that is, as a technology is use for longer, everybody can employ it more productively), but also generation specific (that is, older people will have more experience and can therefore use a technology more efficiently than younger people). Both types of specific knowledge are lost upon adoption and result in adoption being less profitable. Additionally, old people have much more to lose than young people as they stand to lose the experience they have gained with the old technology, which makes the older people particularly resistant to change.
21. Fixed costs are introduced in this section only as they do not qualitatively affect the equilibrium with blocked entry. Without any fixed cost, the number of high-tech firms would tend to infinity under free entry.
22. See Chapter 7, de Groot and Nahuis (1998) and Benassy (1998) for a discussion of these issues in a slightly different context.

APPENDICES

A. The basic model

A.1 The allocation of labour

In this appendix, we solve for the allocation of labour in the basic model. For this aim, we first need to establish consumer and producer behaviour. Each consumer maximizes utility in two steps. In the first step, he maximizes his consumption index subject to his budget constraint (where it leads to no confusion, we suppress indices):

$$\max\ c = x^{\beta} y^{1-\beta} \quad s.t. \quad xP_X + yP_Y \leq cP_c = I\ , \tag{A.1}$$

where x refers to the demand of an individual consumer for a bundle of high-tech commodities (with corresponding price index P_X), y refers to the demand for traditional goods (with corresponding price P_Y), and I is income (which equals wage income $w\bar{l}$ for a worker and profits for an owner of ideas). Our non-satiation assumption implies that the budget constraint is always binding. Notice that profit income does not accrue to workers, but flows back to the idea owners. The corresponding first order conditions are given by $x = \beta I/P_X$ and $y = (1-\beta)I/P_Y$. Or, combining these two conditions,

$$\frac{xP_X}{yP_Y} = \frac{\beta}{1-\beta}\ . \tag{A.2}$$

So the income share allocated to high-tech and traditional goods is equal to β and 1- β, respectively (and is independent of the level of income). The price index P_c is then straightforwardly derived as

$$P_c = \left(\frac{P_X}{\beta}\right)^{\beta}\left(\frac{P_Y}{1-\beta}\right)^{1-\beta}\ . \tag{A.3}$$

In the second step, he maximizes x subject to the income share allocated to high-tech products

$$\max\ x \quad s.t. \quad \sum_{i=1}^{N} \breve{x}_i p_{xi} \leq xP_X \ . \tag{A.4}$$

This yields the individual's demand for good i

$$\breve{x}_i = x\left(\frac{p_{xi}}{P_X}\right)^{-\varepsilon} \quad \text{where} \quad P_X = N^{1-\sigma}\left[\frac{1}{N}\sum_{i=1}^{N} p_{xi}^{1-\varepsilon}\right]^{\frac{1}{1-\varepsilon}} \ . \tag{A.5}$$

Producers in the X-sector maximize profits, $\pi_i = X_i p_{xi} - L_{Xi} w$, subject to the production technology and the demand equation for good i. X_i is the aggregate demand for good i. From the first order condition for a maximum, we arrive at the well-known mark-up relation

$$p_{xi} = \frac{\varepsilon}{\varepsilon - 1}\frac{w}{\mu f_i L_{Xi}^{\mu-1}} \ . \tag{A.6}$$

Firms in the Y-sector operate under perfect competition so

$$P_Y = w \ . \tag{A.7}$$

Under the assumption of symmetry of high-tech goods consumption, we can then derive that

$$x = N^{\sigma}\breve{x} \quad \text{and} \quad P_X = N^{1-\sigma}\frac{\varepsilon}{\varepsilon - 1}\frac{w}{\mu f L_{Xi}^{\mu-1}} \ . \tag{A.8}$$

Due to the utility function being homothetic, it holds that

$$\frac{\breve{x}_i}{y} = \frac{f_i L_{Xi}^{\mu}}{L_Y} \, . \tag{A.9}$$

We can then rewrite goods-market equilibrium as

$$\frac{xP_X}{yP_Y} = \frac{N^{\sigma}\breve{x}_i N^{1-\sigma}\dfrac{\varepsilon}{\varepsilon-1}\dfrac{w}{\mu f_i L_{Xi}^{\mu-1}}}{yw} = \frac{N\varepsilon L_{Xi}}{\mu(\varepsilon-1)L_Y} = \frac{\beta}{1-\beta} \, . \tag{A.10}$$

In combination with the requirement for labour-market equilibrium,

$$L_Y + NL_{Xi} = L \, , \tag{A.11}$$

we arrive at the solution for the allocation of labour to the high-tech sector,

$$L_{Xi} = \frac{\beta\mu(\varepsilon-1)L/N}{(1-\beta)\varepsilon + \beta\mu(\varepsilon-1)} \, , \tag{A.12}$$

which corresponds to equation (5.7) in the text.

A.2 The critical productivity increase

Using that $c = x^{\beta}y^{1-\beta}$ in combination with the above results, we find

$$c = w\frac{\bar{l}}{P_c} = w\frac{\bar{l}}{\left(\dfrac{P_X}{\beta}\right)^{\beta}\left(\dfrac{P_Y}{1-\beta}\right)^{1-\beta}} = \bar{l}\left(\frac{\beta(\varepsilon-1)\mu f L_X^{\mu-1}}{\varepsilon N^{1-\sigma}}\right)^{\beta}(1-\beta)^{1-\beta} \, . \tag{A.13}$$

Note that adoption increases f with a factor γ so we can derive that $c^A = \gamma^\beta c^{NA}$. The F-function for the medium aged is then given by

$$F_{t_1,t_1-1} = \delta[c^A_{t_1,t_1-1} - c^{NA}_{t_1,t_1-1}] - \nu\bar{s} = \delta(\gamma^\beta - 1)c^{NA}_{t_1,t_1-1} - \nu\bar{s} . \qquad \text{(A.14)}$$

The critical productivity increase to keep the middle-aged consumers indifferent between adopting and not adopting is found by putting F equal to zero and is given by

$$\gamma^* = \left[1 + \frac{\nu\bar{s}\left(\varepsilon N^{1-\sigma}\right)^\beta}{\delta\bar{l}\left(\beta(\varepsilon-1)\mu f L_X^{\mu-1}\right)^\beta (1-\beta)^{1-\beta}}\right]^{1/\beta} . \qquad \text{(A.15)}$$

This expression corresponds to equation (5.13) in the text.

A.3 Division of the cake

The next step is to determine the effects of an increase in productivity on the share of the 'total cake' that accrues to workers. This share is equal to the share that workers get in total nominal income and thus equals

$$\frac{wL}{wL + N\pi_i} = \frac{L}{L + NL_{Xi}\left(\dfrac{\varepsilon}{(\varepsilon-1)\mu} - 1\right)} , \qquad \text{(A.16)}$$

where L_{Xi} is determined from equation (A.12). According to this equation, the labour income share increases when competition is intensified.

B. Free entry

B.1 The allocation of labour and the equilibrium number of firms

In this appendix, we will solve the model under the regime of free entry and exit of firms. For this aim, we need to slightly change the producer's problem in that we have to add a fixed cost. We express the fixed cost in terms of labour. The introduction of this fixed cost is essential to obtain a finite solution for the number of high-tech firms, but it does not affect the basic conclusions to be drawn. The *i*-th producer's maximization problem thus looks like

$$\max \ \pi_i = X_i p_{xi} - (L_{Xi} + L_{fi})w \ , \tag{B.1}$$

where L_{fi} denotes firm *i*'s fixed production costs. Since fixed production costs do not affect the firm's optimal pricing strategy, the mark-up relation is again given by equation (A.6). Entry of new high-tech firms continues until profits are driven down to zero (assuming perfect divisibility of firms). Imposing symmetry and putting π_i=0 yields

$$L_X = \frac{\mu(\varepsilon-1)L_f}{\varepsilon - \mu(\varepsilon-1)} \ . \tag{B.2}$$

This is equation (5.15) in the main text. Using the symmetry assumption, labour market equilibrium is determined by

$$L_Y + N(L_X + L_f) = L \ . \tag{B.3}$$

Combination of these expressions gives the number of high-tech producers,

$$N = \frac{\beta L}{L_f}\left[1 - \frac{\mu(\varepsilon-1)}{\varepsilon}\right], \tag{B.4}$$

which corresponds to equation (5.16) in the text.

B.2 The critical productivity increase

Similar to Appendix A.2 we can determine the critical productivity increase and the impact of increased competition. Again:

$$F_{t_1,t_1-1} = \delta[c^{A}_{t_1,t_1-1} - c^{NA}_{t_1,t_1-1}] - \nu\bar{s} = \delta(\gamma^{\beta}-1)c^{NA}_{t_1,t_1-1} - \nu\bar{s} \ . \tag{B.5}$$

Substituting (A.13), (B.2) and (B.4) in the *F*-function and rewriting yields:

$$F_{t_1,t_1-1} = A\delta(\gamma^{\beta}-1)(\varepsilon-1)^{\mu\beta}\varepsilon^{-\sigma\beta}(\varepsilon-(\varepsilon-1)\mu)^{(\sigma-\mu)\beta} - \nu\bar{s} \ , \tag{B.6}$$

where

$$A = L^{(\sigma-1)\beta+1}\beta^{\sigma\beta}\mu^{\beta\mu}f^{\beta}(1-\beta)^{(1-\beta)}L_f^{(\mu-\sigma)\beta} \ . \tag{B.7}$$

The explicit solution for γ^* follows directly from setting (B.6) equal to zero. By straightforward application of the implicit function theorem the sign of $\partial\gamma^*/\partial\varepsilon$ is determined by:

$$-\frac{\left(\frac{\varepsilon}{\varepsilon-1}\mu-\sigma\right)(\varepsilon-(\varepsilon-1)\mu)+(1-\mu)(\sigma-\mu)\varepsilon}{\varepsilon(\varepsilon-(\varepsilon-1)\mu)} \ . \tag{B.8}$$

Since we impose μ to be smaller than $\varepsilon/(\varepsilon-1)$, the sign of the derivative of the critical productivity increase with respect to the elasticity of substitution is determined by the numerator of (B.8). It is then straightforwardly shown that:

$$\text{sgn.}\left(\frac{\partial\gamma^*}{\partial\varepsilon}\right) = \text{sgn.}\left(\sigma-\frac{\varepsilon}{\varepsilon-1}\right) \ . \tag{B.9}$$

6. The skill premium and appropriability: Exploring wage dynamics (II)

1. INTRODUCTION

Wage inequality decreased in the 1970s and has increased since the early 1980s in the US (see Table 6.1). The increase in inequality in the 1980s drew a lot of attention. Four explanations have been proposed: a change in educational attainment,[1] a change in institutions like minimum wage laws; a change in the bias in technological change; and increased trade with low-wage countries. Through a process of elimination of explanations that are not clearly supported by the data, biased technological change is thought to be the dominant explanation for the observed pattern in wage inequality. To explain wage inequality further, given this state of affairs, there are two ways to proceed. The first is to test empirically the claim that a bias in technological change is responsible for the increase in inequality. Krueger (1993) examines whether there exists a technology-related skill premium. Computer use turns out to be directly related to a skill premium.[2] A second route to strengthen the case for a biased-technological-change explanation, which is the default 'explanation', is to explain how a biased-technology shock arises and how it is translated into inequality. Most literature, taking the second route, explains how the bias relates to inequality (for example, Galor and Moav, 1998, and Greenwood and Yorukoglu, 1997).

This chapter extends these contributions by taking the next step of explaining where the bias in technology comes from. We develop a model that predicts that an increase in the supply of educated workers *causes* a bias in technological change which can cause the relative wage of skilled workers to increase.

Table 6.1 Non-production wage-bill and employment share, relative wage and R&D intensity and productivity in the US, 1973–89 [a]

	1973	1977	1981	1989
Non-production wage-bill share	0.337	0.351	0.397	0.414
Non-production employment share	0.246	0.261	0.285	0.303
Non-production/production wage differential	1.55	1.53	1.53	1.62
Supply of high education[b,c]	10.8		16.6	21.5
R&D intensity manufacturing	0.063	0.062	0.077	0.1
Patents per million $ R&D[d]	1.7	1.5	1.1	1

Notes: [a] *Source*: Machin and Van Reenen (1998).
[b] *Source*: OECD Employment Outlook (1993).
[c] Share of college educated, for 1970, 1980 and 1990.
[d] *Source*: Kortum (1993).

Our key assumption is that skilled workers perform tasks that are fundamentally different from those of unskilled workers. While unskilled workers contribute directly to (current) production, skilled workers are employed as non-production workers. The latter contribute to the continuous process of organizational change and improvements within a firm that affect the future productivity of the firm. In this process they build on and further expand the knowledge stock that is already accumulated within the firm. Non-production workers both use and produce new knowledge. In this setup, an increase in the supply of skilled workers gives rise to two effects. There is a direct conventional effect that decreases wages of skilled workers in order to induce firms to absorb the increased supply and expand non-production jobs. However, if firms employ more skilled non-production workers, the demand for knowledge inputs in non-production rises. To generate these inputs, skilled workers themselves should be employed, so demand for skilled labour increases. This second effect counteracts the conventional effect. This is an induced *investment* effect: the expansion of non-production jobs triggers investment in organizational knowledge capital. If sufficiently large, the investment effect may offset the conventional effect. Investment incentives are significant if the cost of capital does not increase too quickly in response to increased investment and if the returns to investment can be sufficiently appropriated. Indeed we find that under these circumstances, the firm is willing to attract skilled labour at a higher wage if educational attainment raises, that is, the demand curve for skilled labour slopes upward.

The degree of appropriability as the key determinant of the skill-premium pattern is endogenous in the extended version of our model. In particular, we distinguish between two types of knowledge investment: first, firms can accumulate knowledge internally (in-house R&D), and, second, they can buy technology in the patent market. Internal knowledge accumulation allows firms to internalize part of the intertemporal knowledge spillovers from research, but this is impossible when taking out patents. We show that at low levels of the supply of skilled labour, patents are the dominant source of technology acquisition, appropriability is weak, and skill premiums are mainly determined by the conventional mechanism. However, with a high supply of skilled labour, firm-specific knowledge becomes dominant and appropriability improves. The induced investment effect dominates the conventional effect and an increase in the supply of skilled labour is matched with an increase in the skill premium. Hence, if skilled labour becomes gradually more abundant, the share of patents in total R&D declines steadily, while the skill premium first decreases and then increases.

Our model replicates the main stylized facts that guide the wage-inequality debate. First, the majority of US industries have, despite the increases in the relative cost of skilled workers, increased the ratio of skilled to unskilled labour (Bound and Johnson, 1992, Berman, Bound and Griliches, 1994, Katz and Murphy, 1992). Table 6.1 shows that the wage-bill share of non-production workers increased from the early 1970s to the late 1980s. The non-production employment share increased in both the 1970s and the 1980s whereas the non-production/production wage ratio fell in the 1970s and increased in the 1980s. As a second stylized fact, it is worth stressing that within-industry increases in the ratio of skilled to unskilled workers overwhelm the between-industry shifts (Bound and Johnson, 1992, Berman, Bound and Griliches, 1994, Katz and Murphy, 1992). Although intersectoral reallocations can be included in our model, we do not want to focus on this aspect. Third, our model replicates the finding that skill upgrading is positively correlated with R&D intensity changes (Machin and Van Reenen, 1998). Finally, our model is consistent with the long-term fall in the number of patents generated per R&D dollar.[3]

This chapter relates to the literature that analyses the interaction between technology, wage inequality and the endowment of skills. There are two strands. The first takes a biased-technology shock as given and explains the effect on wage inequality. Caselli (1999) for example shows how a technology revolution affects inequality if the workforce is heterogenous in training costs. Greenwood and Yorukoglu (1997) analyse how an acceleration in investment-specific technology affects productivity growth and wage inequality if skilled workers have a comparative advantage in technology implementation. These analyses leave unexplained where the technology shock comes from. The second strand of analysis, including Acemoglu (1998), Kiley (1999), Lloyd-Ellis (1999) and Galor and Moav (1998) – where our contribution fits in – explains where the bias comes from. Acemoglu (1998) is closely related to our approach in the sense that increases in the endowment of skills are causal in the increase in inequality. In Acemoglu, similar to the analysis in this chapter, the upward slope of the long-run demand curve for skill is driven by induced investment. The mechanism and assumptions driving the upward slope are, however, markedly different. First, Acemoglu, who treats skilled and unskilled workers as symmetric factors of production, ignores the different nature of production versus non-production work, a distinction we acknowledge explicitly.

Second, we do not rely on such a strong separation between different sources of knowledge accumulation, but focus on the interaction and complementarities between these sources. Third, our model provides a new testable prediction, born by the data, that the number of patents per R&D dollar decreases by increases in the supply of skilled labour. Fourth, Acemoglu introduces a sector bias where we introduce a bias within sectors. Finally, we argue that decreasing returns in the knowledge accumulation process will *ultimately* reverse the prediction that a skill-supply shock will increase wage inequality. Closely related to Acemoglu's analysis is Kiley (1999) who analyses a deterministic version of this model. The induced-innovation mechanism is analogous. Lloyd-Ellis (1999) explicitly separates out the development of new technologies and the absorption of these in production. For both activities skills are important. Depending on the available distribution of skills, technologies can be introduced faster than they are absorbed such that wage inequality is driven up. Galor and Moav (1998), similarly, study the observations discussed earlier but emphasize the endowment of ability and within (skill) group inequality.

The emphasis in this chapter on the investment character of skilled work relates the chapter to the literature showing that skilled labour has a comparative advantage in implementing technology and R&D (Bartel and Lichtenberg, 1987). Furthermore, this chapter uses building blocks from the literature on endogenous growth like, for example, the intentional accumulation of knowledge by profit-maximizing firms. We combine benchmark R&D growth models, where patents are assumed to take care of rent appropriation (see Grossman and Helpman, 1991, and Romer, 1990), with an approach based on firm-specific knowledge. Hence, we assume that innovation also takes place within firms and that firms rely on both in-house R&D and firm-specific knowledge as well as patents developed using the pool of public knowledge (cf. Peretto, 1998 and 1999, Smulders and van de Klundert, 1995 and Thomson and Waldo, 1994). We use the fact that spillovers are not complete and instantaneous; this fact is well documented (see Jaffe, Trajtenberg, Henderson, 1993). We extend the theory of growth based on firm-specific knowledge by broadening the concept of technological change to organizational change (management, and so on).

The plan of this chapter is as follows. Section 2.1 provides an overview of the basic model which is presented in greater detail in Section 2.2 and Section 2.3. Section 2.4 spells out the different effects that drive the skill premium. In the remainder of Section 2, the general-equilibrium impacts of a supply shock to the skill endowment are analysed. Section 3 extends the basic model by introducing growth based on patents along with growth based on firm-specific knowledge. There we endogenize the degree of appropriability, which is crucial for the induced investment mechanism. Section 4 concludes.

2. A GENERAL-EQUILIBRIUM MODEL OF NON-PRODUCTION JOBS

2.1 Overview of the model

There is a continuum of firms, each supplying a unique product variant under monopolistic competition. For notational convenience we normalize the mass of firms to unity. Firms hire two types of labour, labelled skilled (H) and unskilled (L). The supply of both types of labour is exogenously given.[4]

As explained in the introduction, we interpret skilled workers as non-production workers, to be contrasted with production workers who are supposed to be less skilled (note that the empirical literature that documents the change in the skill premium applies the production/non-production classification). Indeed, education and training results in two types of skills. The more elementary skills consist of basic insights and capabilities to undertake given activities at a certain accuracy. Beyond this type of skill, there are more sophisticated skills to analyse existing activities and to generate new knowledge. Those who obtained the first type of education are called unskilled workers. Those who also obtained the second type are called skilled workers. We may think of skilled workers as marketing managers, organization experts, financial planners, research lab workers and so on.

Firms maximize profits and consumers maximize utility. Consumers have Dixit-Stiglitz preferences over a variety of goods. We consider a closed economy without uncertainty with perfect foresight in which all markets clear.

2.2 Preferences and households' behaviour

The consumer side of the model follows the by now standard approach of growth theory. The representative consumer cares about an index (C) of differentiated consumption goods (x_i) à la Dixit and Stiglitz (1977):

$$C = \int_0^1 x_i^{\frac{\varepsilon}{\varepsilon - 1}} di^{\frac{\varepsilon - 1}{\varepsilon}} , \tag{6.1}$$

where ε is the constant elasticity of substitution. Consumers maximize intertemporal welfare that features a constant discount rate (ϑ) and constant elasticity of intertemporal substitution ($1/\rho$):

$$U_t = \int_t^{\infty} e^{-\vartheta t} \frac{C_t^{1-\rho} - 1}{(1-\rho)} dt . \tag{6.2}$$

Maximization of (6.1) and (6.2) subject to the appropriate budget constraints implies that the price elasticity of demand for any good x_i equals ε, and that the change of consumption over time is governed by the Keynes-Ramsey rule:

$$r - \hat{p}_c = \vartheta + \rho \hat{C} , \tag{6.3}$$

where r is the nominal interest rate.

2.3 Technology and firm behaviour

Each firm produces according to the following production function:

$$x_i = f_i^{\beta} L_i^{\delta} \equiv F(f_i, L_i) \ . \tag{6.4}$$

Final output is denoted x. L are unskilled workers[5] whose effectiveness or productivity depends on knowledge or (organizational) quality (f) created by skilled workers. Skilled workers (H) gradually improve the organization, production technology or, for example, the perceived quality (by means of marketing). The one-dimensional variable f captures all these different aspects. The stylized representation of the accumulation process for ideas is as follows:

$$\dot{f}_i = \xi f_i^{1-\alpha} \bar{f}^{\varphi} H_{fi}^{\lambda} \equiv G(f_i, \bar{f}, H_{fi}), \quad \alpha, \varphi \in [0,1), \ \lambda \in (0,1] \ . \tag{6.5}$$

The (non-production) work of skilled workers improves the production environment for the unskilled workers.[6] The useable knowledge stock consists of two types of knowledge inputs. First, skilled workers analyse, exploit and expand the stock of accumulated firm-specific experience and organizational knowledge capital (f_i).[7] Second, skilled workers benefit from knowledge developed in other firms, as captured by the average knowledge level, $\bar{f}$, in other firms. This second knowledge input is beyond the control of the individual firm and gives rise to an intertemporal knowledge-spillover externality that is familiar from R&D-based endogenous growth models. However, firms internalize the intertemporal spillover effect from their own knowledge generation to their own non-production activities: they take into account that accumulation of specific knowledge not only affects production but also provides inputs for future research. Firms do not internalize the intertemporal spillovers to other firms because they cannot appropriate the associated returns. We assume that there are non-increasing returns to scale with respect to both types of knowledge taken together. That is, we either assume constant returns to scale ($\varphi = \alpha$), which implies endogenous growth (Romer, 1990), or diminishing returns to scale ($\varphi < \alpha$), which implies semi-endogenous growth (Jones, 1995b).

Keely and Quah (1998) argue that a review of the empirical literature on R&D, technology and growth shows that: (1) inputs in the knowledge production function are strongly related to knowledge output and (2) output of knowledge production is inaccurately proxied by patents, as 'Most knowledge accumulation does *not* occur from private firms' R&D producing *patentable* knowledge.'[8] This latter observation motivates our emphasis on knowledge as a firm-specific asset.[9] A final observation is that spillovers do occur, but do not happen automatically or completely. Hence, we do not assume *perfect* nor automatic knowledge spillovers, as is clear from the distinction between $\bar{f}$ and f_i in our specification.

Novel here is the interpretation that the regularities related to spillovers and knowledge accumulation – familiar to the R&D-based endogenous growth literature – are extended to all non-production activities. To see the analogy between R&D and other non-production work think of a new way of organizing a firm. The implementation and development of new organizational schemes often takes years and builds on past experience. From the organizational scheme that a specific firm works out more general principles can be useful for other firms too. If this information is written down or disseminated in some way, other firms might benefit too ($\bar{f}$). However, the next reorganizing firm might use this information but still needs to go through the process of convincing, motivating and adapting to its specific 'own' circumstances[10] (that is increasing the firm's specific knowledge stock, f_i). Though we argue the model to be applicable to the broad category of all non-production workers, the remainder of the analysis is expressed in R&D terms only.

Demand for unskilled and skilled workers is determined by the firms' production and non-production decisions, given that firms are restricted by a downward-sloping demand curve for final output.

Skilled workers' output is not sold in the market directly.[11] By hiring skilled workers, the firm can invest in firm-specific knowledge. The wage the firm is willing to pay for skilled labour depends thus on the return to investment in firm-specific knowledge. The firm invests up to the point that the marginal return to investment in firm-specific knowledge equals the cost of capital. This no-arbitrage condition can be written as:[12]

$$\frac{p\left(1-\frac{1}{\varepsilon}\right)\frac{\partial x}{\partial f}+q\frac{\partial \dot{f}}{\partial f}+\dot{q}}{q}=r\ , \tag{6.6}$$

where ε is the elasticity of demand, r is the cost of capital and q is the shadow value of firm-specific knowledge, or in other words the firm's internal accounting price for non-production workers' output. The left-hand side of the equation represents the marginal return from non-production workers' activities. The first term represents the value of their contribution to improving production efficiency (the accumulation of productive knowledge), the second term represents the value of their contribution to providing knowledge inputs for the process of investment in firm-specific knowledge (the expansion of useable knowledge), and the final term captures a capital gain (capturing the fact that if the price to knowledge accumulation increases over time, investing today in knowledge becomes more attractive).

The firm hires skilled labour up to the point where the marginal cost of hiring (the wage for skilled labour, w_H) equals its marginal product which is the marginal amount of knowledge it generates $\partial \dot{f}/\partial H_f$ valued at the price of knowledge q:

$$w_H = q\frac{\partial \dot{f}}{\partial H_f}\ . \tag{6.7}$$

Similarly, the firm hires unskilled labour up to the point where the marginal cost of hiring (the wage for unskilled labour, w_L) equals its marginal revenue product. Taking into account that the firm's demand curve for final goods slopes downward we find:

$$w_L = p\left(1 - \frac{1}{\varepsilon}\right)\frac{\partial x}{\partial L} . \tag{6.8}$$

2.4 Short-run partial equilibrium

We now reduce the conditions governing firm behaviour into one equation. We assume that firms are symmetric. Because the mass of firms is one, in equilibrium the representative firm absorbs all labour (L_i=L and H_{fi}=H_f=H). The symmetry assumption also implies that f and $\bar{f}$ grow at a common rate, denoted by g, which can be written (from (6.5)) as:

$$g = \xi f^{-(\alpha-\varphi)} H_f^{\lambda} . \tag{6.9}$$

Assuming that both H and L grow at a common rate l and evaluating the partial derivatives, we find:

$$\frac{w_L}{w_H}\frac{L}{H_f^{1-\lambda}}\frac{\beta\lambda\xi}{\delta}f^{\varphi-\alpha} + (\beta-\varphi)g + (\hat{w}_H - \hat{w}_L) + (\delta-\lambda)l = r - \hat{p} , \tag{6.10}$$

where hats denote growth rates. This equation equates the real cost of capital to the real return to investment. We can use it to study the partial-equilibrium short-run effects of an increase in the supply of high-skilled labour H by keeping the knowledge level f and the real cost of capital r-$\hat{p}$ constant. Whenever the return to investment increases (the l.h.s. of (6.10) goes up), there will be an induced demand for skilled labour and hence an upward pressure on their relative wage.

The first term on thc l.h.s. represents the *direct* effect of investment for production. If more skilled labour is employed, the first term becomes smaller.[13] The reason is that the costs of productivity improvements, relative to production costs, rise with H because of diminishing returns ($\lambda<1$). Hence, a higher supply of skilled labour reduces their relative wage if firms only care about their ability to improve productivity in production (that is, if only the first term on the l.h.s. matters). This is the *conventional effect* of an increase in skills.

The other terms on the l.h.s. reflect the effects of knowledge growth on the future relative costs of (or return to) organizational change. Let us first consider the term $(\beta-\varphi)g$. On the one hand, the larger the impact of organizational change, as captured by g, on productivity of unskilled labour – as captured by β – the more attractive it is to invest. Large increases in productivity induce firms to invest now rather than later. To be able to undertake these investments, firms increase their demand for skilled labour. On the other hand, if knowledge spillovers from other firms are large, knowledge growth results in large reductions in the cost of organizational change over time, which reduces the incentive to invest now and makes firms willing to postpone investment, thus it reduces the demand for skilled labour. If the scope for productivity improvements (β) is large and spillovers (φ) are small, firms are willing to invest more in response to a higher growth rate. Ceteris paribus, this increases demand for skilled labour. Employing more skilled labour increases the rate of productivity growth. Hence, if spillovers are small, this may increase the demand for skills. Thus the demand curve for skilled labour tends to slope upward because of this channel.

The third term on the l.h.s., the rate of increase in the skill premium, reflects the fact that if skilled labour becomes more expensive to hire over time, investment becomes more expensive over time, and it is attractive to undertake investment now rather than in future. This increases the demand for skills, ceteris paribus.

The fourth term on the l.h.s. is a net scale effect. As long as population growth is constant, this term is constant. An increase in l increases the return to investment today if decreasing returns in research set in less strongly than those in production.

2.5 General equilibrium

We now close the model by taking into account goods-market equilibrium and capital-market equilibrium. The former implies $c=x$ and $p_c=p$. The capital market is in equilibrium if the rate of return satisfies the Keynes-Ramsey rule (6.3), which can now be written as $r-\hat{p}=\vartheta+\rho\hat{x}$. Combining (6.3), (6.9) and (6.10), and using (6.4) to solve for $\hat{x}$, we find:

$$\left(\rho-(1-\varphi/\beta)-\frac{L}{H_f}\frac{\lambda}{\delta}\frac{1}{w_H/w_L}\right)\beta g+\vartheta_l=\widehat{w_H/w_L}\,, \qquad (6.11)$$

where $\vartheta_l\equiv\vartheta+[\lambda+\delta(\rho-1)]l$. This equation in two unknowns, the skill premium w_H/w_L and the growth rate g, subsumes equilibrium in the markets for output, unskilled labour and capital and will be shortly referred to as the no-arbitrage condition. By confronting it to the condition for equilibrium in the skilled labour market, equation (6.9), we can determine the dynamics in general equilibrium.

2.6 The upward-sloping demand curve and endogenous growth

We first illustrate the main result of an upward-sloping demand curve for skilled labour in a special case of the model. We assume that there are constant returns with respect to knowledge accumulation in the non-production activities, that is $\varphi=\alpha$. As a result, the rate of growth in the economy depends on the supply of skilled labour only; see (6.9). To avoid accelerating growth rates, we assume that there is no population growth ($l=0$); the usual assumption in endogenous growth literature.

The model is now fully represented by equation (6.9) and (6.11). Figure 6.1 depicts equation (6.9) as the vertical line labelled GG. The SS-curve in the figure is the locus for which the skill premium w_H/w_L is constant as can be derived from equation (6.11). Full employment of skilled labour requires that the economy is always on the GG line. The skill premium jumps immediately to its long-run value, given by the point of intersection between the GG line and the SS curve. That is, the economy immediately jumps to the steady state.

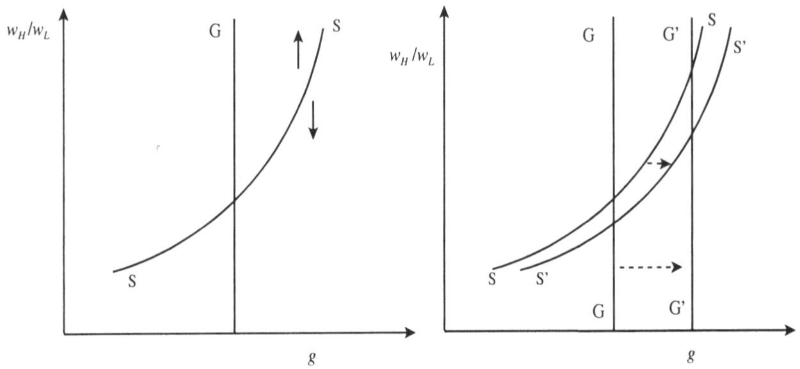

Figure 6.1 Endogenous growth

We are now ready to analyse the general equilibrium consequences of an increase of the supply of skilled labour for wage inequality. If H increases, the SS-locus shifts down and the GG-line shifts to the right. Hence the change in the skill premium cannot be unambiguously determined from the figure. Fortunately, we can easily find a closed form solution for the skill premium. Substituting (6.9) into (6.11), and taking into account that $\varphi=\alpha$ and that the skill premium is constant, we find:[14]

$$w_H/w_L = \frac{(\lambda/\delta)L}{(\vartheta/\beta\xi)H_f^{1-\lambda} + [\rho-(1-\varphi/\beta)]H_f} . \tag{6.12}$$

Differentiation with respect to H_f reveals that the condition for a rise in the skill premium is:

$$1-\alpha > \rho\beta + (1-\beta) + (1-\lambda)\frac{\vartheta}{g} . \tag{6.13}$$

This last condition neatly reveals four determinants that may cause the demand curve for skills to slope upward.

First, appropriability of the (intertemporal) returns to non-production activities (as measured by 1-α) should be high. The mirror image of this is that spillovers to other firms (as measured by φ) should be low. This underlines our key assumption that skilled workers create the knowledge that is subsequently used as an essential input in non-production activities. If new knowledge only affects the firm's production activities and all knowledge inputs in non-production activities come from outside (that is α=1), condition (6.13) could never be satisfied and the demand curve for skills slopes conventionally downward. Note that most of the endogenous growth literature considers this case by assuming that all intertemporal spillovers from research are external effects for the individual firm.

Second, the cost of capital should not rise too fast with increased investment. Note from (6.3) that ρ governs the sensitivity of interest rates with respect to investment. This emphasizes the importance of the assumption that non-production labour is engaged in the *investment* process, rather than the production process.[15] If firms hire more skilled labour, investment and growth rises in the economy, forcing households to save more. This induces them to require a higher rate of return on their savings, especially when they prefer a smooth consumption pattern (ρ large). When firms face a higher cost of capital, investments in firm-specific knowledge by hiring more skilled labour, becomes less attractive. The rise in the cost of capital thus mitigates the demand for skilled labour. For the demand curve for skills to slope upward, the cost of capital should not increase too much, which happens if the intertemporal substitution elasticity is high (1/ρ high).

Third, diminishing returns with respect to knowledge in production (as measured by $1-\beta$) reduce the skill premium. Diminishing returns reduce the value of additions to the knowledge stock, which are generated by hiring more skilled labour.[16]

Fourth, diminishing returns with respect to skilled labour in non-production activities (as measured by $1-\lambda$) reduce the skill premium.

Diminishing returns reduce the marginal value of skilled labour for the firm and thus depress their wage.

To summarize, an upward-sloping demand curve for skills requires that the intertemporal returns from an expansion of non-production activities accrue mainly to the firm's skilled workers rather than to shareholders (in the form of higher rates of return) or other firms (because of spillovers). Moreover, the returns should not fall too quickly because of diminishing returns in production or non-production activities. Only under these circumstances, the firm passes through the intertemporal returns to non-production workers in the form of higher wages for skilled labour.

2.7 The upward-sloping demand curve and non-scale growth

We now turn to the more general version of the model and assume that there are diminishing, rather than constant, returns with respect to knowledge in non-production activities, and we take into account population growth ($\varphi<\alpha$, $l>0$). The main difference with the case in the previous section is that now long-run growth becomes independent of the size of the skilled labour force. Hence, there is no scale-effect on the growth rate from an increase in the supply of skills. We show that the upward-sloping demand curve arises under fairly similar conditions, but vanishes in the long run.

The growth rate depends on the supply of skilled labour and on the stock of firm-specific knowledge accumulated in the past, see equation (6.9). Accordingly, the growth rate is a predetermined variable that changes over time. Differentiating (6.9) with respect to time yields the equation of motion for the growth rate:

$$\hat{g} = \lambda l - (\alpha - \varphi) g \ . \tag{6.14}$$

Hence the GG locus for a constant growth rate reads

$$g = \lambda l \ / \ (\alpha - \varphi) \ . \tag{6.15}$$

The SS-locus is the same as in the case of endogenous growth[17] and follows directly from (6.11). Figure 6.2 depicts the phase diagram that results from equations (6.11) and (6.14). Transitional dynamics occur along the upward-sloping saddle path.

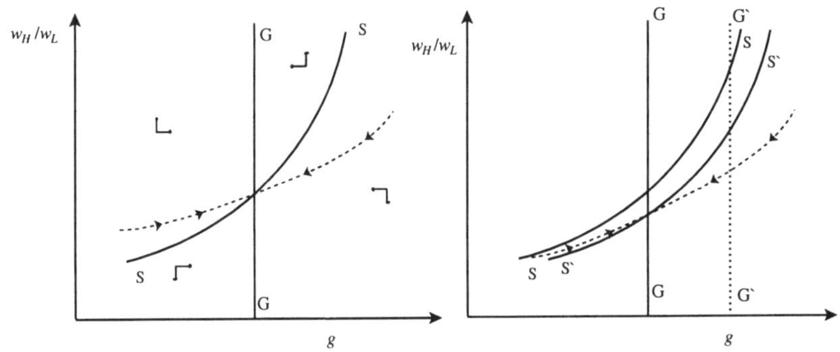

Figure 6.2 Non-scale growth

To analyse the consequences of an increase in the supply of skilled labour, we now need to distinguish between long-run and short-run effects. For simplicity, we consider a permanent increase in H at t=0, but allow H to grow at rate l at all other dates.

The long-run growth rate is not affected by the supply shock (the GG-locus remains unchanged), while the SS-curve shifts downward. Hence in the long run, the skill premium unambiguously declines in response to an increased supply of skilled labour.

An increase in skilled labour increases the demand for knowledge inputs in non-production activities on impact. Indeed, the expansion of non-production activities results in an increase in the stock of firm-specific knowledge in the long run (or to be more precise, an increase in knowledge per skilled worker). But, because of diminishing returns in non-production activities with respect to knowledge, this causes the returns to additional knowledge inputs to fall, thus offsetting the increase in the demand for the skilled labour that produces these knowledge inputs. In the long run, increases in firm-specific knowledge cannot make *non*-production workers more productive because of diminishing returns. Hence, in the long run, there is no increase in demand for skilled labour for this purpose. In the long run knowledge only makes *production* more productive, and putting in more effort in this sphere requires a fall in the skill premium (the conventional effect).

In the short run, however, the growth rate increases by the expansion of non-production jobs. The combination of the shift of the SS locus and the short-run increase of the growth rate produces a (short-run) result that is very similar to that in the endogenous growth case. Indeed, the skill premium may increase in the short run.[18] The difference with the endogenous growth case is that now the (short-run) equilibrium is not determined by the intersection of the SS locus but by the saddle path, which is below the SS locus, thus making an increase in the skill premium less likely. The reason is that firms anticipate a decline in the skill premium (since in the long run the skill premium decreases), that is, they anticipate a decline in the cost of non-production activities. This gives firms an incentive to postpone non-production activities and to reduce the demand for skilled labour. This partly reverses a possible increase in the skill premium.

To derive an exact condition for the upward-sloping demand curve to arise, we linearize equations (6.11) and (6.14) around the steady state and calculate the short-run response of the skill premium to a change in the supply of skilled labour. The linearized system reads:

$$\begin{bmatrix} \dot{\widetilde{w_H/w_L}} \\ \dot{\tilde{g}} \end{bmatrix} = \begin{bmatrix} a+\vartheta_l & -\vartheta_l \\ 0 & -\lambda l \end{bmatrix} \begin{bmatrix} \widetilde{w_H/w_L} \\ \tilde{g} \end{bmatrix} + \begin{bmatrix} a+\vartheta_l \\ 0 \end{bmatrix} \tilde{H}_f , \tag{6.16}$$

where tildes refer to percentage deviations from the initial steady state (log-linearized variables) and $a \equiv \beta g[\rho - (1-\varphi/\beta)]$. From (6.9) we find the initial change (a time t=0) in the growth rate (which is predetermined):

$$\tilde{g}(0) = \lambda \tilde{H}_f \,, \tag{6.17}$$

where $\tilde{H}_f$ is the permanent shock to the skill endowment. The stable root of this system is λl. Hence, we can calculate the jump in the skill premium as:

$$w_H \tilde{/} w_L(0) = -\left(\frac{\lambda l + a + (1-\lambda)\vartheta_l}{\lambda l + a + \vartheta_l} \right) \tilde{H}_f \,. \tag{6.18}$$

The skill premium increases in the short run if the expression in parentheses is negative. Taking into account the definition of *a* given above, we find the following condition:

$$1-\alpha > \rho + (1-\beta)\frac{\varphi}{\beta} + (1-\lambda)\frac{\vartheta_l}{\beta g} \,. \tag{6.19}$$

Note that this is exactly the same condition as for the endogenous growth case, see (6.13), although now $\varphi<\alpha$ and $1-\alpha$ and φ no longer sum up to unity (and ϑ takes a different value because of population growth). Hence, the very same mechanisms as already explained in the previous case apply.

3. APPROPRIABILITY

The model discussed in the previous section can explain the upsurge in inequality in the 1980s provided that the intertemporal returns can be appropriated sufficiently. To be able to explain *both* the decrease and the increase in inequality in the 1970s and 1980s respectively, the degree of appropriability should have altered over time. Two questions arise immediately. First, is there any evidence that the degree of appropriability changed? If so, then how and why did the degree of appropriablity change?

This section addresses these questions. We do so in reverse order, as we need to pin down more precisely what appropriability is before the evidence can be judged.

3.1 A model with endogenous appropriable knowledge

In the model above we have distinguished between three types of knowledge which are jointly produced in a single type of research activity. The three types are (i) knowledge applied in the production activities of the firm that developed the knowledge (productive knowledge), (ii) knowledge that is used in the research activities of the firm that developed the knowledge (useable knowledge), and (iii) knowledge that can be freely used in the research activities of firms that did not develop it. The fewer spillovers of the latter type there are, the better firms can appropriate the rents that their research gives rise to. The degree of appropriability is likely to depend on the mode of organization of research.[19] Firm-specific research is aimed at specific knowledge tailored to the firms' circumstances and thus less useful for other firms. Moreover, in-house organization of R&D facilitates secrecy of research results.

In this section we not only consider firm-specific research activities, but extend the analysis for research activities that are undertaken by research units independent of production firms and that yield knowledge that is less specific to firms. The knowledge output is sold in a patent market. Appropriability in this research activity is smaller than for firm-specific research. In particular, we assume that a new patent improves production of firms, for which the inventor is fully rewarded, but it also improves productivity of research for which the inventor is not rewarded. Trade in ideas among firms requires a patent register in which technical details and general principles are documented and by which they become freely available to other researchers. As in almost all R&D-based growth models (Romer, 1990, Grossman and Helpman, 1991 as well as almost all models in Aghion and Howitt, 1998a), in our patent sector, researchers build on the total stock of public knowledge, but cannot internalize the contribution they make to this stock.

Arbitrage in the capital market ensures that both research types yield the same return. However, shocks to the economy – in particular a change in the supply of skilled labour – endogenously change the relative importance of both research sectors and therefore shift the economy-wide degree of appropriability.

A formalization

Two types of knowledge positively affect the productivity of unskilled labour in final-goods production:

$$x_i = f_i^{\beta} n_i^{1-\beta} L_i \,, \tag{6.20}$$

where n is the number of patents a firm has purchased.[20] Patents are produced in an R&D sector where firms may enter freely. A firm that devotes H_n units of skilled labour to research produces $\chi \bar{n}^{\mu} \bar{f}^{\psi}$ patents. The productivity of research firms is increasing in both types of knowledge distinguished. The production function a patent firm faces thus is:

$$\dot{n} = \chi \bar{n}^{\mu} \bar{f}^{\psi} H_n \,. \tag{6.21}$$

The public knowledge stock ($\bar{n}^{\mu} \bar{f}^{\psi}$) consists of registered patents $\bar{n}$ and spillover from firm-specific knowledge. Free entry of patent firms implies that:

$$p_n = \frac{w_H}{\chi \bar{n}^{\mu} \bar{f}^{\psi}} \,. \tag{6.22}$$

Thus, the price of a patent, p_n, equals the production cost. The other ingredients of the model are essentially as in the previous section except that we allow for spillovers of the patent firms to the production of firm-specific knowledge.[21] Hence:

$$\dot{f}=\xi f^{1-\alpha}\bar{f}^{\varphi}\bar{n}^{\nu}H_f\ . \tag{6.23}$$

The demand for patents follows from the no-arbitrage condition analogous to equation (6.6):[22]

$$\frac{p\left(1-\dfrac{1}{\varepsilon}\right)\dfrac{\partial x}{\partial n}+\dot{p}_n}{p_n}=r\ . \tag{6.24}$$

Furthermore, firms decide on production and labour for research. Production or demand for unskilled labour is governed by (6.8) and the hiring of skilled labour by (6.7) and (6.6). Comparing equation (6.24) with (6.6) reveals the crucial difference between the two types of research: the private return to firm-specific research includes a term valuing the contribution of current research to future research productivity ($q\partial\dot{f}/\partial f$), while the private return to developing patents does not include such an intertemporal return. Equilibrium on the market for skilled labour now reads: $H=H_f+H_n$.

The remainder of this section discusses symmetric steady-state equilibria with endogenous growth. The endogenous-growth requirement implies the following parameter restrictions: $\varphi=\alpha-\nu$ and $\psi=1-\mu$. Before proceeding, some additional notation is introduced. Denote $A\equiv f/n$. Thus, the ratio of the stock of firm-specific knowledge to the number of patents is defined as A. This ratio is an indication of the degree of appropriability in the economy's research-capital stock. We can rewrite (6.23) and (6.21) as $\hat{f}=\xi A^{-\nu}H_f$ and $\hat{n}=\chi A^{\psi}H_n$ to get the following expression for balanced growth:

$$g=\frac{\chi A^{\psi}H}{1+(\chi/\xi)A^{\nu+\psi}}\ . \tag{6.25}$$

Equation (6.25) is depicted in the upper panel of Figure 6.3 as the balanced growth (BG)-curve.

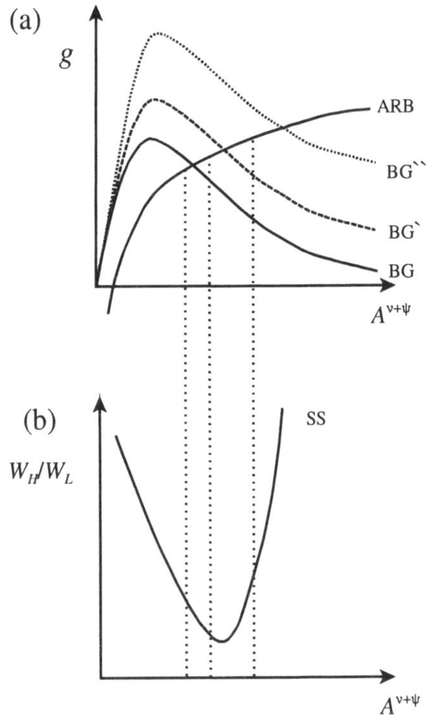

Figure 6.3 General equilibrium with endogenous appropriability

The BG-curve is hump-shaped, reaching its maximum at $A^{\nu+\psi}=\psi/\nu$. At a low degree of appropriability (A), the economy has a relative abundance of patents which implies a low productivity of the patent sector. Balanced growth, however, requires both the stock of patents and the stock of firm-specific capital to grow at equal rates. This implies that skilled labour must be largely employed in the low-productive patent sector. Hence growth is low. Increases in A make the patent sector more productive and hence less resource consuming for balanced growth. This allows for a higher growth rate. Increasing A further reduces the research productivity within firms, making the accumulation of firm-specific capital the bottleneck (and lowering the feasible balanced growth rate again). The cost of capital should equal the return to patent development:

$$r-\hat{p} = (1-\beta)\frac{L}{w_H/w_L}\chi A^{\psi} . \tag{6.26}$$

The return to a patent is increasing in the patent elasticity in production, production size (L) and in the reciprocal of the cost-benefit term (the cost of producing a patent and the benefit of saving on unskilled workers' wages). The cost of capital should equal the return to firm-specific knowledge investment:

$$r-\hat{p} = \beta\frac{L}{w_H/w_L}\xi A^{-\nu}+(1-\alpha)g . \tag{6.27}$$

The structure is analogous to (6.26) again, but for the term that indicates the dynamic externality that is appropriated (the strength of this mechanism is governed by 1-α). Using these two expressions with the Ramsey rule, (6.3), and $\hat{x}=g$ from (6.20), the no-arbitrage condition reads:

$$g = \frac{(1-\beta)\chi A^{\nu+\psi} - \beta\xi}{(1-\alpha-\rho)(1-\beta)\chi A^{\nu+\psi} + \rho\beta\xi}\vartheta . \tag{6.28}$$

The upper-panel of Figure 6.3 depicts equation (6.28) as the ARB-curve (for the case where 1-α>ρ).[23] The upward slope, implying that a higher growth rate is to be met with greater scarceness of patents to prevent arbitrage opportunities, is explained by the fact that the return to investment in f is increasing in g (see equation (6.27)). To equalize returns, A has to increase, as is obvious from equation (6.26).

The lower panel depicts the skill premium as a function of A; the degree of appropriability in the economy's research-capital stock:[24]

$$\frac{w_H}{w_L} = \left[(1-\alpha-\rho)(1-\beta)\chi A^{\nu+\psi} + \rho\beta\xi\right]\frac{L}{(1-\alpha)\vartheta A^{\nu}}. \tag{6.29}$$

The skill premium depends negatively on A at low levels of A and positively at high levels of A. Equation (6.29) is depicted as the SS-curve in the lower panel of Figure 6.3 and has an U-shape (again for the case where $1-\alpha>\rho$).

Increasing the supply of skilled labour

An increase in the supply of skilled labour shifts the BG-curve upward to BG′. The intersection of the BG′- and ARB-curve determines the equilibrium that is characterized by an increased degree of appropriability of the research-capital stock. This is consistent with the decline in the number of patents per R&D dollar in the data. In the lower panel we can read that the skill premium decreased (a pattern consistent with facts in the 1970s). Shifting the BG′-curve upward, by increasing the supply of skill again, shows that the degree of appropriability increases further, but now the skill premium increases (a pattern consistent with facts in the 1980s).

Why does the skill premium fall when skilled labour is scarce but rise in a skilled-labour-abundant economy? More skilled labour implies higher growth and larger benefits to internalizing intertemporal spillovers from R&D. The faster growth, the more skilled labour is employed in firm-specific research departments, since firms can appropriate the intertemporal benefits of high growth rates. Higher growth rates increase the demand for skilled labour, thus counteracting the conventional downward pressure on wage premiums. Indeed, equations (6.26) and (6.27) show that higher growth benefits firm-specific R&D; see equation (6.27). Arbitrage shifts the economy towards relatively more firm-specific knowledge: equality of the returns in (6.26) and (6.27) then requires a rise in A. This shift reduces the returns from increasing the firm's labour productivity (see the first term on r.h.s. of (6.27)), and the return to firm-specific R&D relies relatively more on the appropriated intertemporal spillovers (see the second term on the r.h.s. of (6.27)). On the economy-wide level, appropriability of dynamic gains becomes more important when growth increases. Hence, while for low supply of skilled labour and correspondingly low growth rates the conventional effect dominates, for high supply and high growth the appropriability effect dominates, which increases skill premiums.

3.2 The patent-productivity puzzle

Empirical research documents a fall in productivity of R&D in terms of patent output per real dollar of R&D. The fall is found for both the 1970s and the 1980s, that is, a monotonic fall that contrasts with the U-shaped pattern for the skill premium in the same period.[25]

Our extended model not only generates the observed pattern for the skill premium, but also predicts that it is accompanied by a fall in patent productivity. A gradual rise in the supply of high-skilled labour shifts research activity towards more firm-specific research. Typically, firm-specific research generates less visible research output: the possibility of secrecy and the tacitness of the knowledge generated in this way makes the propensity to patent typically lower. Hence, innovation is underestimated in the innovation statistics. For direct evidence that firms indeed rely more on secrecy relative to patents, see Cohen, Nelson and Walsh (2000). As a result, research *output* statistics tend to report a fall in output when research shifts to firm-specific research, because these statistics concentrate on patents. On the research *input* side, however, it is difficult to separate out the inputs in firm-specific research from those aimed at developing patents. Thus, typically, measured patent output falls, but measured input is not corrected for the reduction in inputs directed at patent development.

In the model, the following ratio comes closest to the statistic that is used in the empirical literature on research productivity:

$$\frac{\dot{n}}{Hw_H/p_n}, \qquad (6.30)$$

that is, new patents divided by the total real cost of R&D, ignoring the distinction between inputs into firm-specific research and those into other research. If inputs were measured correctly, the productivity statistic would be $\dot{n}p_n/H_n w_H$, which would be constant and equal to unity due to our assumption of zero profits in the research sector; see (6.21) and (6.22).

However, the ratio above has total inputs H instead of H_n in the denominator, and because of zero profits the ratio boils down to H_n/H which is directly related to the appropriability measure, A. Appropriability, A, is in the steady state (combine (6.25) and $\dot{n}/n = g = \chi A^{\psi}$ from (6.21)):

$$\frac{\dot{n}}{Hw_H/p_n} = \left(\frac{\dot{n}p_n}{H_n w_H}\right)\frac{H_n}{H} = \frac{H_n}{H} = \frac{1}{1+(\chi/\xi)A^{\nu+\psi}} . \qquad (6.31)$$

Hence, when H increases, A increases monotonically, and measured patent productivity falls monotonically. Thus from the 1970s to the late 1980s we register a fall in patent productivity.

4. CONCLUSION

Wage inequality increased in the 1980s in the majority of OECD countries. In the evaluation of the potential explanations for this phenomenon, trade, institutions, technology and relative factor supplies (education), the technology explanation was left by default.

We argue that for this explanation to be convincing, it should be shown what type of technology shock is responsible for the increase in inequality and where the shock comes from. Our explanation, based on explicit recognition of the difference between production and non-production jobs, is consistent with the following empirical observations. From the early 1970s to the late 1980s the relative supply of skilled versus unskilled workers increased, the wage-bill share of non-production workers increased, the ratio of non-production/production workers increased and the productivity of research has fallen. We show that these observations are consistent with a fall in the relative wage of non-production workers in the 1970s and an increased relative wage in the 1980s.

We showed that an increased endowment of skilled labour might induce an increase in the relative wage for skilled labour. The argument that this chapter developed starts with the explicit recognition that skilled labour or non-production workers perform tasks that are similar to investment activities. That is skilled workers produce firm-specific capital. Once it is recognized that skilled workers use knowledge while producing knowledge, an upward-sloping demand curve for skill follows, provided that (i) the degree of appropriability of investment in organizational capital is sufficiently large, (ii) the investment costs do not rise too fast and (iii) diminishing returns related to knowledge accumulation do not set in too strongly.

NOTES

1. Educational attainment clearly increased (see Table 6.1); as such this implies a downward pressure on the relative wage of skilled workers. An alternative way of interpreting the evidence, however, is that the quality of high schools decreased (and hence the quality of unskilled workers decreased). This might put a downward pressure on the relative wage of unskilled workers.
2. This is not uncontroversial. DiNardo and Pischke (1997) argue that the relation is spurious. However, Autor, Katz and Krueger (1998) reconfirm earlier findings of Krueger (1993).
3. Though not apparent from Table 6.1, in the late 1980s the number of patents per R&D dollar increased again. We do not focus on this aspect here as it is still unclear how important the numerous institutional changes with respect to the patent system are in explaining this (see Kortum and Lerner, 1998 and Jaffe, 1999).
4. An extension of the model by allowing for endogenous skill formation is not pursued here as it does not add to the point made in this chapter. Heterogenous learning ability would be necessary to avoid a corner solution with only skilled workers.
5. Note that we allow for decreasing returns to unskilled labour $0<\delta<1$. The underlying assumption is that the firm also employs a fixed factor whose size is normalized to one.
6. There are decreasing returns with respect to skilled labour inputs. This captures the 'stepping on toes effect', indicating congestion and duplication in research (see Jones, 1995b, for an extensive discussion).
7. For an extensive discussion on the firm-specific nature of knowledge, see Smulders and van de Klundert (1995) and Peretto (1999). For an explicit treatment of the tacitness of knowledge, see Dosi (1988).
8. See page 3 of their paper, second italics added.
9. We will explore the interaction between firm-specific and patentable knowledge in Section 3.

10. Jovanovic (1997) argues that adjustment and implementation costs of ideas dominate the non-rivalness of knowledge.
11. Think of output of the marketing manager, the human-resource manager, the product designers and the organizational experts.
12. The following three equations can be straightforwardly derived from the firm's dynamic optimization problem. Firms maximize profits, discounted by interest rate r, subject to (6.4),(6.5) and the downward-sloping demand curve for its output. Suppressing the firm index i, we may write the Hamiltonian as $p(F(f,L))\cdot F(f,L)-w_L L-w_H H_f+qG(f,\bar{f},H_f)$, where $F()$ is the production function in (6.4) and $G()$ is the accumulation function in (6.5).
13. If more *un*skilled labour is employed (L higher), the return to investment is higher since more production workers benefit from the same increase in productivity due to the non-rivalness of knowledge. Hence, an increase in unskilled labour raises the demand for skilled labour to undertake the investments which leads to an increase in the skill premium.
14. Remark that the results are expressed in terms of the skill premium, w_H/w_L, in this chapter, instead of ω ($\equiv w_L/w_H$) as in Chapters 3 and 4.
15. The results, however, are not affected by allowing skilled labour in final output production too. As long as substitution between skilled and unskilled labour is imperfect, the results stand upright. See Appendix A.
16. This effect vanishes if there are no spillovers (i.e. if φ=0). In this case, all knowledge is created inside the firm with constant returns with respect to knowledge (see (6.5) and note that α=φ=0). On the firm level a constant rate of growth of knowledge can be attained equal to ξH_f^λ which translates into a constant rate of total factor productivity growth equal to $\beta\xi H_f^\lambda$. Hence, diminishing returns are no longer important. In contrast, if the firm relies on outside knowledge (φ>0), it takes into account that an increase in firm-specific investment *reduces* its rate of knowledge growth for a given rate of growth in the outside knowledge stock, $\bar{f}$. This hurts the firm more, the more important the diminishing returns in the use of knowledge in production are (as measured by $1-\beta$).
17. Except for the fact that ϑ_1 takes a different value because of population growth.
18. In the short run – before the per-capita stock of knowledge has fully adjusted – non-production workers become more productive if more knowledge inputs are available. This gives firms the incentive to hire more labour to produce this knowledge.
19. The degree of appropriability is defined as the fraction of rents associated with a certain new idea that accrues to the firm that developed the idea.
20. We now set δ equal to unity to avoid cluttering the analysis with symbols. δ is easily introduced without affecting the main results.
21. We suppress the firm index. Recall that there is a mass of monopolistically competitive firms of size one.
22. Appendix B provides a step-by-step derivation of the model in this section.
23. Appendix B discusses the case where $1-\alpha<\rho$.

24. Use the two expressions for the returns to investment, (6.26) and (6.27) with the Ramsey rule (6.3).
25. The upsurge in patenting (even per R&D dollar) in the late 1980s is, according to Kortum and Lerner (1998), associated with an increase in research productivity that could be mimicked in the model by increasing the exogenous research productivity in the patent sector. This point is further ignored as it is empirically still unclear how important the institutional changes from 1984 onward have been in affecting the patent practice (see Jaffe, 1999).

APPENDICES

A. Skilled labour in production

In the main text we assumed that all skilled labour is allocated to non-production work with an investment character. Though most non-production work is well captured by this perspective, some skilled workers' jobs are more closely related to production. One might, for example, think of supervision labour and the skilled workers who implement and monitor newly developed quality strategies. In this appendix we explore the robustness of the upward-sloping demand curve for the assumption that some skilled work is closely related to production and hence does not have an investment character. We do so by extending the model of Section 2 by allowing for skilled labour in production.

A.1 Production-related skilled work

The production function for final output is now denoted as: $F(f_{xi}, L_i)$. The effective organizational quality, f_{xi}, depends now on the mass of skilled workers in activities related directly to production. Hence:

$$f_{xi} = G_1(f_i, H_{xi}) \ . \qquad \text{(A.1)}$$

where H_x is skilled labour allocated to production-related activities.[1] Analogously equation (6.5) is now formulated as:

$$\dot{f}_i = G_2(f_i, \bar{f}, H_{fi}) = \xi f_i^{1-\alpha} \bar{f}^{\varphi} H_{fi}^{\lambda} \ . \qquad \text{(A.2)}$$

H_f is skilled labour in investment-related activities. We assume that substituting (A.1) in (6.4) results in the following CES production function:[2]

$$x = Af^{\beta}\left[\frac{\delta_L}{\delta_L+\delta_H}L^{\frac{\sigma_{HL}-1}{\sigma_{HL}}}+\frac{\delta_H}{\delta_L+\delta_H}H_x^{\frac{\sigma_{HL}-1}{\sigma_{HL}}}\right]^{\frac{\sigma_{HL}}{\sigma_{HL}-1}(\delta_L+\delta_H)}, \tag{A.3}$$

where firm subscripts are suppressed. Next, the model economy is analysed in the special case that generates endogenous growth, $\varphi=\alpha$.

A.2 The demand for skill and capital-market equilibrium

The allocation of skilled labour in production adds another first-order condition for the solution of the firm's optimization problem. Firms allocate skilled workers to production-related activities up to the point where the marginal cost of hiring (the wage of skilled workers) equals the marginal revenue. This condition, divided by equation (6.8), looks like:

$$\frac{L}{H_x}=\left(\frac{\delta_L}{\delta_H}\frac{w_H}{w_L}\right)^{\sigma_{HL}}. \tag{A.4}$$

This condition shows more explicitly the *conventional effect* of changes in factor endowments. An increase in the endowment of skilled labour is met by a decrease in the skill premium. Ample substitution possibilities (σ_{HL} large) ease the absorption of skilled workers and hence mitigate the necessary skill premium decrease. Using (A.4), (6.9) and the resource constraint, $H_f=H-H_x$, the growth rate can be formulated as:

$$g=\xi\left(H-L\left(\frac{w_H}{w_L}\frac{\delta_L}{\delta_H}\right)^{-\sigma_{HL}}\right)^{\lambda}. \tag{A.5}$$

Remember that we analyse the endogenous growth version of our model, hence $\varphi=\alpha$. The growth rate is increasing in the skill premium. A higher relative wage for skilled workers reduces their attractiveness for production-related activities and hence leaves more skilled workers for growth-enhancing investment activities. Combining this equation with the no-arbitrage condition (6.6) yields, after some manipulation, an implicit equation for the relative wage:

$$\frac{w_H}{w_L} = \frac{\dfrac{1}{\delta_L(\delta_L+\delta_H)} L\left[\delta_L+\delta_H\left(\dfrac{w_H}{w_L}\dfrac{\delta_L}{\delta_H}\right)^{1-\sigma_{HL}}\right]}{(\vartheta/\beta\xi) + [\rho-(1-\varphi/\beta)]\left[H-L\left(\dfrac{w_H}{w_L}\dfrac{\delta_L}{\delta_H}\right)^{-\sigma_{HL}}\right]} . \qquad \text{(A.6)}$$

This is an analogous expression to equation (6.12) where λ is set to unity for simplicity. Note that in both the numerator and the denominator a reallocation effect is present. The numerator shows that a higher relative wage affects the value of the production of final output whereas the denominator shows that reallocation of skilled labour affects the value of organizational capital.

Close inspection (or careful application of the implicit-function theorem) shows that the sign of the expression in brackets before H, that is the sign of $\rho-(1-\varphi/\beta) = \rho+(1-\beta)\varphi/\beta-(1-\alpha)$, determines the sign of the effect of an increase in H on the skill premium. Hence, we find a similar condition as in the model in the main text (see equation (6.13) with $\lambda=1$).

B. Solution of the model in Section 3

B.1 Firms

The current-value Hamiltonian of the optimal control problem for firm i reads as:

$$\mathcal{H}_i = p_i x_i - w_H H_{fi} - w_L L_i + q_{fi}\xi f_i^{1-\alpha}\bar{f}^{\varphi}\bar{n}^{\nu}H_{fi} + q_{ni}I_{ni} - p_{ni}I_{ni} . \qquad \text{(B.1)}$$

Use equation (6.20), (6.23) and the demand curve for an individual product variety to arrive at the following first-order conditions:

$$\frac{\partial \mathcal{H}_i}{\partial L_i}=0 \Leftrightarrow p_i=\left(\frac{\varepsilon-1}{\varepsilon}\right)\frac{w_L}{f_i^{\beta}n_i^{1-\beta}}, \tag{B.2}$$

$$\frac{\partial \mathcal{H}_i}{\partial H_{fi}}=0 \Leftrightarrow q_{fi}f_i \leq \frac{w_H}{\xi f_i^{1-\alpha}\bar{f}^{\varphi}\bar{n}^{\nu}} \text{ ,with equality if } H_{fi}>0 , \tag{B.3}$$

$$\begin{aligned} &\frac{\partial \mathcal{H}_i}{\partial f_i}=rq_{fi}-\dot{q}_{fi} \Leftrightarrow \\ &p_i\left(\frac{\varepsilon-1}{\varepsilon}\right)\beta f_i^{\beta-1}n_i^{1-\beta}L_i+(1-\alpha)q_{fi}\xi f_i^{-\alpha}\bar{f}^{\varphi}\bar{n}^{\nu}H_{fi}=rq_{fi}-\dot{q}_{fi} , \end{aligned} \tag{B.4}$$

$$\frac{\partial \mathcal{H}_i}{\partial I_{ni}}=0 \Leftrightarrow q_{ni}\leq p_{ni} \text{ ,with equality if } I_{ni}>0 , \tag{B.5}$$

$$\frac{\partial \mathcal{H}_i}{\partial n_i}=rq_{ni}-\dot{q}_{ni} \Leftrightarrow p_i\left(\frac{\varepsilon-1}{\varepsilon}\right)(1-\beta)f_i^{\beta}n_i^{-\beta}L_i=rq_{ni}-\dot{q}_{ni} . \tag{B.6}$$

These expressions, with two transversality conditions, determine optimal firm behaviour. The two labour-market constraints, $H=H_{fi}+H_n$ and $L=L_i$, free entry for firms in the patents sector (6.22) and the Ramsey rule (6.3)

complete the description of the economy. We are interested in interior, symmetric equilibria. This implies that (B.5) and (B.3) hold with equality and $p_i = p \wedge f_i = \bar{f} \wedge n_i = \bar{n}$ for all i. It is easy to show that with $\varphi = \alpha - \nu$ and $\psi = 1 - \mu$ positive balanced growth is possible and implies that $\hat{f} = \hat{n} = \hat{w}_H - \hat{p} = \hat{w}_L - \hat{p} \equiv g$ and $\hat{p} = \hat{q}_n = \hat{q}_f = \hat{p}_n$.
Balanced growth requires

$$\frac{\xi f^{1-\nu} n^{\nu}}{f} H_f = \frac{\chi \bar{n}^{1-\psi} \bar{f}^{\psi}}{n} (H - H_f) \ , \tag{B.7}$$

where the labour-market equilibrium for skilled workers and the parameter restrictions for endogenous growth are used. Use $A \equiv f/n$ to write

$$H_f = \frac{\chi A^{\psi} H}{\xi A^{-\nu} + \chi A^{\psi}} \ , \tag{B.8}$$

and substitute this in $g = \xi A^{-\nu} H_f$ to obtain the expression for the BG-curve, equation (6.25) in the main text.

Research arbitrage says that the returns to both research activities are equalized to the return consumers require on their savings (equation (6.3)). Use (6.22), (B.6) and (B.2) to get the return for buying patents

$$r - \hat{p} = (1 - \beta) \frac{L}{w_H / w_L} \chi A^{\psi} \ . \tag{B.9}$$

Analogously, the return for investing in appropriable knowledge f can be rewritten as (use (B.2), (B.4) and (B.3)):

$$r - \hat{p} = \beta \frac{L}{w_H / w_L} \xi A^{-\nu} + (1 - \alpha) g \ . \tag{B.10}$$

The equality of both investment options can be used to derive an expression for w_H / w_L in terms of g and A. Substitute this in (B.9) and set this offered return equal to the required return from the Ramsey rule (6.3) to get the arbitrage condition (equation (6.28) in the main text). This curve slopes downward if 1–α<ρ (see, Figure B.1); if 1–α>ρ holds, the curve also slopes U shaped.

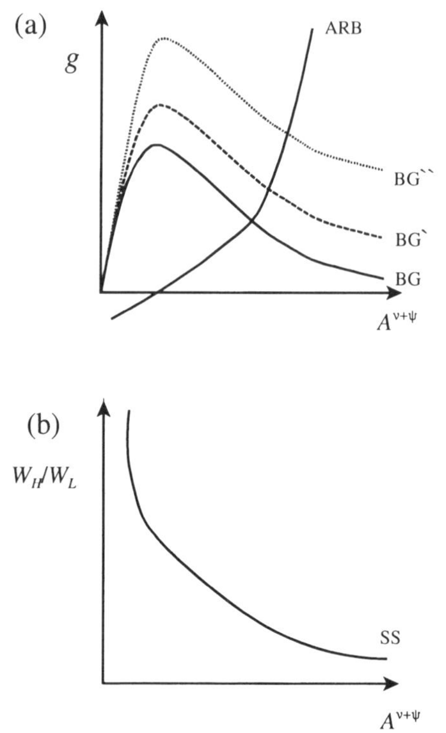

Figure B.1 General equilibrium in the model of Section 3 when ρ>1-α

NOTES

1. In the main text f_i and f_{xi} were assumed identical.
2. With σ_{HL} equal to unity we arrive at the Cobb-Douglas production function: $x=Af^{\beta}L^{\delta_L}H_x^{\delta_H}$.

7. Specific technology, variety, spillovers and welfare

1. INTRODUCTION

Product variety is an important determinant of economic welfare. Following the seminal work of Dixit and Stiglitz (1977) and Spence (1976), the welfare effects of variety have been analysed from various angles.[1] Dixit and Stiglitz themselves conceive the problem of optimal diversity (in a static context) as one of quantity versus diversity. With the presence of economies of scale in production, producing a limited variety of goods saves resources that can be used to extend the production volume. Hence a trade-off arises that gives rise to the question of social optimality of the market equilibrium. It turns out that the market supports too low product diversity.

Subsequent studies addressed the optimality question in the presence of growth. In a dynamic context, reduced variety saves resources that can be used not only to extend the production quantity, but potentially also to increase the rate of growth. Grossman and Helpman (1991, Chapter 3) analyse welfare in a model of endogenous growth. In their analysis, there is (continuous) growth in product variety resulting from investment in R&D. The more labour an economy allocates in the R&D sector, the less labour remains for production of consumption goods. The question here is one of growth in variety versus volume of consumption goods. The optimal trajectory entails more rapid growth of variety than the market equilibrium sustains, as firms ignore the contribution of their knowledge creation to a common 'knowledge pool'. Grossman and Helpman (1991) also analyse a quality ladder model. In this model quality is endogenous and variety is exogenous. Here innovative effort aimed at quality improvement might be sub-optimally high or low, depending on the size of the quality step. Van de Klundert and Smulders (1997) develop an endogenous growth model in which R&D is an in-house activity aimed at improving quality. Besides quality growth, variety is also determined endogenously. An important characteristic of the model is the constancy of variety in equilibrium. The authors analyse the welfare consequences of different regimes of competition in the presence of knowledge spillovers. They find sub-optimally low growth. Product variety can be both too small or too large.

The studies discussed so far assume that variety has a direct effect on consumers' welfare as consumers have a love of variety. Another branch of literature looks at the productivity effects of increased product variety (Ethier, 1982). Romer (1990) takes this route and develops a model in which diversity of capital inputs grows at a constant rate, thereby increasing productivity at a constant rate. In a similar spirit, the focus in this chapter will be on the effects of diversity on productivity. Barro and Sala-i-Martin (1995, Chapter 6) point to the similarity of endogenous growth models of, on the one hand, an expanding variety of consumption goods directly adding to utility, and, on the other hand, the models of expanding variety of producer goods that increase productivity and add indirectly to utility. Ultimately, it is the flow of utility that matters. Still, they argue strongly in favour of expanding product variety models that add to productivity as they can more easily be reconciled with reality.

This chapter develops a two sector model with endogenous growth to shed light on the advantages and disadvantages of using a greater or more limited variety of differentiated inputs. In the model *both* the growth rate of product quality and the number of product varieties is endogenous (Smulders and van de Klundert, 1995 and Peretto, 1996 and 1999 apply a similar model). In contrast to the existing literature that studies the tension between growth and variety, we explicitly single out returns to variety (that are external to the firms producing the brands) from the elasticity of substitution between brands.[2] In, for example, Dixit and Stiglitz (1977), Grossman and Helpman (1991) and van de Klundert and Smulders (1997), the returns to variety and the elasticity of substitution between brands are mechanically linked. Although the returns to variety and the elasticity of substitution are likely to be (negatively) related, there is no reason to assume a link as tight as suggested in the above-mentioned papers. As will be shown in this chapter, disentangling this relation is crucial for the welfare effects that are derived. In other contexts, not characterized by both endogenous (quality) growth and endogenous variety, the importance of separating returns to variety from the elasticity of substitution is convincingly shown, see for example, Benassy (1996 and 1998), Broer and Heijdra (1996), de Groot and Nahuis (1998) and Holtz-Eakin and Lovely (1994). These studies put welfare results, derived in the context of models using the traditional Dixit-Stiglitz specification for returns to variety, in perspective.

The model we construct has two sectors. In the first sector of the model, the assembly sector, firms produce a homogeneous consumption good. Production takes place using (homogeneous) labour that is supplied inelastically and differentiated intermediate goods. The intermediates are produced in the second sector of the model. In this sector, firms are producing a unique brand of an intermediate good. Labour is employed in each firm for production and for research. Intentional investments in R&D aimed at increasing quality in this sector drive economic growth. R&D is modelled as an in-house activity yielding completely firm-specific knowledge. Hence, the relevant knowledge base for R&D is the stock of knowledge that is built up by the firm's own past research activities. This contrasts strongly with growth models that assume an intertemporal knowledge spillover that has a public good character and hence creates a distortion (Romer, 1990, Grossman and Helpman, 1991 and Aghion and Howitt, 1998a). Note, therefore, that the familiar intertemporal spillover, in our model, is non-distortionary.

The reasons emphaszing the analysis on a model with completely firm-specific knowledge and thus without an intertemporal externality are twofold. First, there is convincing evidence that an important part of knowledge is tacit (hence knowledge has a firm-specific character), see Dosi (1988). Second, and more important, most R&D concerns the rival activity of knowledge adoption. That is, every firm has to be engaged in costly 're-invention' and adjustment to the specific circumstances of any idea from the public knowledge pool before it can be used efficiently. This aspect of knowledge accumulation is a factor of 20 to 30 more important in terms of costs than the creation of non-rival ideas, according to Jovanovic (1997), who forcefully makes a this point. We incorporate this perspective, and take it to the opposite extreme of the benchmark endogenous growth models, in that R&D is (a rival) in-house process for firms. This departure from the public knowledge pool concept is of course crucial from a welfare point of view. At the end of the chapter we introduce a knowledge pool that is partially public – that is we allow for spillovers – to examine the robustness of our findings.

Distortions that *are* present are the following. The uniqueness of brands gives rise to market power. The resulting non-competitive pricing of differentiated products results in a distortion as it affects the relative price of inputs in production of the final consumption good. Two other potentially distorting market failures are present. The first arises because entrepreneurs do not take into account the surplus of their entry decision that accrues to the producers of the consumption good, due to increasing returns to variety. The second can be traced to the fact that entrants ignore the fact that entry implies that the total market has to be shared by more firms. The first will be dubbed the *product diversity effect* and the second will be called, following Mankiw and Whinston (1986), the *business stealing effect*.[3]

We show that explicitly separating out these three potential distortions is crucial for understanding the results of the welfare analysis. Crucial is the strength of the product diversity effect relative to the elasticity of substitution. The market results in growth that is too high and a sub-optimally low level of variety when the product diversity effect is strong. If the business stealing effect dominates (and hence the returns to variety are relatively low), the market growth rate is too low. If the diversity effect is extremely weak, the variety supplied will be sub-optimally high.[4] If it is extremely strong, the opposite holds. In the intermediate case, *both* the rate of growth and variety can be too low. Decentralization of the first best solution is shown to be possible and to require two instruments, namely an output subsidy on the production of intermediates and a lump-sum tax or transfer to producers of intermediates.

Hence, we contribute to the analyses of optimal growth by analysing optimal government policy in a model where growth and variety are endogenous and intertemporal spillovers are absent. Moreover, returns to variety are distinguished from the elasticity of substitution. Next we add to the literature on second-best policy by considering the optimal choice of the number of permits to be issued to intermediate goods producers (hence, we analyse optimal policy when the government only has one instrument at its disposal, namely setting variety). The optimal policy in this second-best setting is to refrain from using the number of permits to manipulate the growth rate.

This chapter is organized as follows. In Section 2, we will discuss and present the basic model. The steady state equilibrium of the market economy is presented in Section 3. In Section 4, we perform a first best analysis by looking at the unconstrained social optimum, and we compare this with the market outcome. In addition, we design a policy that supports the first best outcome as a competitive equilibrium. Section 5 considers the second best problem of choosing the welfare maximizing number of firms (by issuing permits), taking as given the non-competitive behaviour after entry. In Section 6 knowledge spillovers are introduced. We end with conclusions and an evaluation in Section 7.

2. THE MODEL

Our economy comprises two sectors. The assembly sector produces a homogeneous consumption good using intermediates and labour. Firms in this sector operate under perfect competition and take the prices of intermediates and wages as given. The intermediates are imperfect substitutes in production of the homogeneous consumption goods. Following Ethier (1982), there is an externality in this sector in the form of increasing returns to variety (productivity increases with variety). In the intermediate goods sector, N firms are operating. These firms compete monopolistically. We assume N to be sufficiently large so that competition is monopolistic à la Chamberlin.[5] The number of intermediate goods producers is determined endogenously by a process of entry or exit as long as profits are non-zero.[6] Intentional R&D performed by firms in this sector results in productivity increases and thus positively affects the quality of intermediates. We assume R&D to be an in-house activity. There is no spillover of the fruits of R&D whatsoever; an assumption that will be relaxed in Section 6. Next, we will describe in turn consumer behaviour, the assembly sector and the intermediate goods sector.

2.1 Consumer behaviour

A representative consumer maximizes his intertemporal utility subject to a dynamic budget constraint[7]

$$\max \int_{o}^{\infty} u(C_t) e^{-\theta t} dt \quad s.t. \quad \dot{W}_t = r_t W_t + w_t L_t - C_t P_{Ct}, \tag{7.1}$$

where C_t is a consumption index, θ is the subjective discount rate, W_t are assets possessed by consumers,[8] r_t is the interest rate, $w_t L_t$ is wage income in the economy and $C_t P_{Ct}$ is expenditures on consumption goods. We assume a constant working population L. Assuming a CRRA utility function, that is $u(C_t) = C_t^{1-\rho}-1/(1-\rho)$, we arrive at the Ramsey rule

$$\frac{\dot{C}_t}{C_t} = \frac{1}{\rho}\left(r_t - \frac{\dot{P}_{Ct}}{P_{Ct}} - \theta \right), \tag{7.2}$$

where $1/\rho$ is the intertemporal elasticity of substitution in consumption. Consumers prefer a steeper consumption profile, the larger the gap between the real rate of interest $(r_t - \dot{P}_{Ct}/P_{Ct})$ and the subjective discount rate (θ), and the larger the intertemporal elasticity of substitution (1/ρ).

2.2 Assembly sector

The assembly sector produces final consumption goods C under perfect competition. The goods are produced according to (dropping time indices where it leads to no confusion)

$$C = X^{\beta} L_C^{1-\beta} \quad \text{with} \quad X = N^{\sigma}\left[\frac{1}{N}\sum_{i=1}^{N} x_i^{\frac{\varepsilon-1}{\varepsilon}}\right]^{\frac{\varepsilon}{\varepsilon-1}}. \tag{7.3}$$

In this formulation, L_C represents the number of production workers in the assembly sector. X is a composite of the intermediates that are available. This specification of the composite is borrowed from Broer and Heijdra (1996). N represents the number of varieties of the intermediate good (indexed i) available and ε is the elasticity of substitution between any pair of differentiated intermediates. Returns to variety are represented by the parameter σ. With all the x_i equal to a common x, as will be the case in equilibrium, we get $X=N^{\sigma-1}(Nx)$. Suppose that there are two bundles of intermediates that are equally large ($N^1x^1=N^2x^2$). If σ>1, the bundle with the largest variety (the largest N) is most productive (that is, there are returns to variety). Parameter restrictions σ≥1 and ε>1 ensure that production exhibits returns to variety (with strict inequality) and that every variety is demanded. The returns to variety equal the special value assumed by Dixit and Stiglitz (1977) if σ=ε/(ε-1). The composite of intermediates specified in equation (7.3), can hence be seen as a generalization of Dixit and Stiglitz (1977).[9]

The producers take prices for intermediates, output prices and wages as given and perform the following (two stage) maximization problem

$$\max_{L_C, X} \Pi_C = CP_C - wL_C - P_X X, \tag{7.4}$$

where Π_C denotes profits and P_X is the price index for the composite good.

Optimization yields

$$\frac{\partial \Pi_C}{\partial X} = 0 \;\Leftrightarrow\; \beta C P_C = X P_X, \text{ and} \tag{7.5}$$

$$\frac{\partial \Pi_C}{\partial L_C} = 0 \;\Leftrightarrow\; (1 - \beta) C P_C = L_C w, \tag{7.6}$$

which is the standard Cobb-Douglas result of fixed income shares.

In the second stage, the firm decides on the optimal amount of input of each variety. The following optimization problem is solved

$$\operatorname*{Max}_{x_i} \; N^{\sigma}\left[\frac{1}{N}\sum_{i=1}^{N} x_i^{\frac{\varepsilon-1}{\varepsilon}}\right]^{\frac{\varepsilon}{\varepsilon-1}} \quad \text{s.t.} \quad \sum_{i=1}^{N} x_i p_{xi} = X P_X, \tag{7.7}$$

which results in

$$\begin{aligned} x_i &= N^{(\varepsilon-1)(\sigma-1)-1}\left[\frac{p_{xi}}{P_X}\right]^{-\varepsilon} X, \\ &\text{where } P_X \equiv \frac{\sum_{i=1}^{N} x_i p_{xi}}{X} = N^{1-\sigma}\left[\frac{1}{N}\sum_{i=1}^{N} p_{xi}^{1-\varepsilon}\right]^{\frac{1}{1-\varepsilon}}, \end{aligned} \tag{7.8}$$

which is the downward sloping demand curve for an intermediate of brand *i*. It is easily seen that under symmetry the true price index ($P_X = N^{1-\sigma}p_x$) is decreasing in *N* if $\sigma > 1$.

2.3 The intermediate goods sector

The intermediate goods sector consists of N firms, each producing a brand of a differentiated input used in the assembly sector. Firm i produces a quality-adjusted amount x_i of the intermediate of type/brand i according to

$$x_i = f_i L_{xi}, \tag{7.9}$$

where L_{xi} represents production labour in the intermediate goods sector, and f_i is labour productivity. The quality of final goods production can be affected by R&D. The stock of knowledge, which is assumed to be completely firm-specific, determines the quality level directly. Knowledge accumulates according to

$$\dot{f}_i = \xi f_i L_{ri}, \tag{7.10}$$

where L_{ri} represents research labour and ξ is the research productivity parameter. The firms aim at maximizing their present discounted value subject to the demand for intermediates and the costs and benefits of engaging in R&D. The objective can hence be written as

$$\max \int_0^\infty \left[x_i p_{xi} - (L_{xi} + L_{ri} + L_f) w\right] e^{-rt} dt. \tag{7.11}$$

L_f is fixed labour required to start production, and r is the interest rate at which firms could invest their money in the financial market.

The current-value Hamiltonian corresponding to this dynamic optimization problem reads as

$$H = x_i p_{xi} - (L_{xi} + L_{ri} + L_f) w + q_{fi} \xi f_i L_{ri}, \tag{7.12}$$

where q_{fi} is the shadow price of knowledge.

The first-order conditions of the intertemporal optimization problem are (assuming symmetry so that we can drop the brand index i)

$$\frac{\partial H}{\partial L_x} = fp_x(1-\frac{1}{\varepsilon}) - w = 0 \quad \Leftrightarrow \quad p_x = \frac{\varepsilon}{\varepsilon-1}\frac{w}{f}, \tag{7.13}$$

according to which firms engage in mark-up pricing (note that the mark up ($\varepsilon/(\varepsilon$-1)) reflects the returns to variety in case $\sigma=\varepsilon/(\varepsilon-1)$, see footnote ?),

$$\frac{\partial H}{\partial L_r} = -w + p_h \xi h = 0 \quad \Leftrightarrow \quad w = p_h \xi h, \tag{7.14}$$

showing that firms allocate R&D labour to this sector as long as the marginal benefits of doing so ($q_f \xi f$) exceed the marginal costs (w), and

$$\begin{aligned} -L_x p_x \frac{\varepsilon-1}{\varepsilon} - q_f \xi L_r &= \dot{q}_f - r q_f \quad \Leftrightarrow \\ \frac{\dot{q}_f}{q_f} + L_x \frac{p_x}{q_f}\frac{\varepsilon-1}{\varepsilon} + \xi L_r &= r, \end{aligned} \tag{7.15}$$

which is the no-arbitrage condition. Investing an amount q_f in the financial market at the rate r should yield the same return as investing in knowledge capital which yields a capital gain, an increase in production, and an increase in the knowledge base.

The model is closed by imposing full employment labour market equilibrium

$$L = L_C + N(L_x + L_r + L_f), \tag{7.16}$$

and imposing instantaneous profits in the intermediate goods sector (π) to equal zero

$$\pi = xp_x - (L_x + L_r + L_f)w = 0. \tag{7.17}$$

Characterization of the steady-state solution of the model will be the topic of the next section.

3. SOLUTION OF THE MODEL

In this section, we characterize the *steady-state* solution of the model and discuss its main comparative statics. We assume that excess profits (or losses) are competed away by free entry and/or exit of firms (ignoring integer constraints). The number of intermediate goods producers is thus determined endogenously. We define the growth rate of labour productivity in the intermediate goods sector as $g\,(\equiv \dot{f}/f)$. In the remainder of this chapter, we take the wage rate as numéraire ($w = 1$).

Solving the model (see Appendix A for a derivation of the expressions) yields a required and planned rate of return on savings and investment, respectively

$$r = \theta + \beta(\rho - 1)g \quad \text{and} \quad r = (\varepsilon - 1)(g + \xi L_f). \tag{7.18}$$

The required real rate of rate of return on savings $(r - \dot{P}_C/P_C = r + \beta g)$ thus increases with the rate of growth of consumption (βg) due to the wish of consumers to smooth their consumption over time ($\rho>0$). The realized rate of return depends positively on the rate of growth, the productivity of research and the fixed cost requirement. Confronting the realized and required rates of return results in a savings-investment equilibrium and yields the equilibrium growth and interest rates[10]

$$g = \frac{(\varepsilon-1)\xi L_f - \theta}{\beta(\rho-1)-(\varepsilon-1)} \quad \text{and} \quad r = (\varepsilon-1)\frac{\beta(\rho-1)\xi L_f - \theta}{\beta(\rho-1)-(\varepsilon-1)}. \tag{7.19}$$

Finally, we derive the equilibrium number of intermediate goods producers (the variety of intermediates) and the allocation of labour as

$$N = \frac{[\beta(\rho-1) - (\varepsilon-1)]\xi\beta L}{\varepsilon[\beta(\rho-1)\xi L_f - \theta]} \quad \text{and} \quad L_C = (1-\beta)L. \tag{7.20}$$

An important feature of the model is that the scale of the economy (L) leaves the growth rate unaffected (see van de Klundert and Smulders (1997)), whereas it leads to an equi-proportionate change in the number of firms. The reason is that an increase in the size of the economy leaves individual firm size unaffected and thus leaves the incentive to engage in R&D unchanged. Scale effects are thus present only in the sense that productivity *levels* are affected. There are, in other words, static returns to scale.[11] Another important notion is that the parameter capturing the 'returns to diversity' does not show up in the solution for the equilibrium variety. The market does not take into account the externality resulting from diversity.

For the growth rate, we can conclude that it depends positively on the fixed cost requirement L_f, the elasticity of substitution ε, and the research productivity parameter ξ. An increase in the fixed cost requirement lowers the equilibrium number of firms and increases the market share of each individual firm. This increases the incentive to engage in R&D and consequently the growth rate. A similar type of argument holds for the elasticity of substitution. A large elasticity of substitution lowers the room for firms to make positive profits and thus increases individual firm size. An increase in θ and ρ reduces the consumers' incentive to save and thus increases the costs of acquiring financial means to finance investment in knowledge capital. Firms will respond by investing less in knowledge capital, reducing the rate of growth (and the equilibrium interest rate). The reduction in the investment, required to keep up with competitors, increases the firm's profits, which will lead to entry. An increase in β (the share of intermediates in the production of final consumption goods) increases the required rate of return on savings (as consumption growth increases for a given level of productivity growth). The equilibrium interest and growth rates consequently go down. This will be accompanied by a downscaling by intermediate goods producers of both their research and their production departments. At the same time, the assembly sector will substitute away from labour and towards intermediates (L_C /NL_x decreases). This combination of reduced high-tech firm size and increased demand for intermediates implies an increase in the number of intermediate firms.

4. OPTIMAL PRODUCT VARIETY

In this section, we will derive the first-best social optimum (FBSO) by solving the social planner problem. The social planner maximizes inter-temporal utility of the representative agent subject solely to technical constraints (that is, technology and resource availability). Characterization of the social optimum will be the topic of Section 4.1. In Section 4.2, we will compare the market equilibrium with the social optimum. The results with respect to the optimal product variety, derived under the Dixit-Stiglitz assumption that $\sigma=\varepsilon/(\varepsilon-1)$, turn out not to generalize when this assumption is dropped. This points to the importance of disentangling the returns to diversity and the degree of imperfect substitutability resulting in market power of monopolistic competitors (see also Benassy, 1996, and Broer and Heijdra, 1996). In Section 4.3, we design a policy that supports the first best outcome as a competitive equilibrium.

4.1 The first-best social optimum

In this section, we will look at the first-best optimum in which a social planner is assumed to maximize the utility of the representative agent subject to the accumulation function of firm specific knowledge, the production technology,[12] and the labour market constraint. Hence,

$$\max U=\int_0^\infty \frac{C^{1-\rho}-1}{1-\rho}e^{-\theta t}dt, \tag{7.21}$$

subject to equations (7.3), (7.10), and (7.16). The current-value Hamiltonian corresponding to this optimal control problem is[13]

$$H = \frac{1}{1-\rho}\Big[\big(L - N(L_x + L_r + L_f)\big)^{1-\beta}\big(N^{\sigma} f L_x\big)^{\beta}\Big]^{1-\rho} + q_f \xi f L_r. \tag{7.22}$$

The first order conditions corresponding to this problem are

$$\frac{\partial H}{\partial L_x} = 0 \iff (1-\beta)NL_x = \beta L_C, \tag{7.23}$$

$$\frac{\partial H}{\partial L_r} = 0 \iff q_f = (1-\beta)\frac{C^{1-\rho}N}{L_C \xi f}, \tag{7.24}$$

$$\frac{\partial H}{\partial N} = 0 \iff L_C = \frac{(1-\beta)L}{\beta\sigma + 1 - \beta} = L_C^{FB}, \tag{7.25}$$

$$\dot{q}_f + \frac{\partial H}{\partial f} = \theta q_f \iff \dot{q}_f + \frac{C^{1-\rho}\beta}{f} + q_f \xi L_r = \theta q_f. \tag{7.26}$$

From equation (7.25), it is evident that employment in the assembly sector is constant. Using equation (7.23), it then follows that NL_x is constant. Substitution of the expressions for NL_x and L_C in the labour market constraint (equation (7.16)) gives an expression for L_r and consequently for the optimal growth rate

$$g^{FB} = \xi L_r^{FB} = \frac{\xi\beta(\sigma-1)}{\beta(\sigma-1)+1}\frac{L}{N^{FB}} - \xi L_f. \tag{7.27}$$

The optimal growth rate thus depends negatively on the number of intermediate goods available in the economy. The optimal number of intermediate goods producers can be derived as[14]

$$N^{FB} = \frac{\beta(\sigma-1)(\rho-1)-1}{\beta(\rho-1)\xi L_f - \theta}\,\frac{\beta\xi L}{\beta(\sigma-1)+1} \quad \text{so}$$
$$g^{FB} = \frac{\xi L_f - \theta(\sigma-1)}{\beta(\sigma-1)(\rho-1)-1}. \tag{7.28}$$

We can derive

$$\frac{\partial N^{FB}}{\partial\sigma} > 0;\ \frac{\partial N^{FB}}{\partial\theta} > 0;\ \frac{\partial N^{FB}}{\partial L_f} < 0;\ \frac{\partial N^{FB}}{\partial\xi} < 0;$$
$$\frac{\partial N^{FB}}{\partial L} > 0;\ \frac{\partial N^{FB}}{\partial\rho} > 0;\ \frac{\partial N^{FB}}{\partial\beta} > 0. \tag{7.29}$$

An increase in the returns to variety positively affects the optimal number of differentiated inputs, which is an intuitively clear result. An increase in the subjective discount rate increases the optimal number of varieties of the intermediate good. The intuition behind this result is as follows. A higher discount rate increases the value attached to current production. This higher production can be achieved by increasing the number of varieties used in the assembly sector, as the positive productivity effect of an additional variety outweighs the negative growth effect that results from increased fixed costs in the economy. Increases in the fixed cost decrease the optimal product variety. The increased fixed cost tends to lower the optimal rate of growth (if N^{FB} is kept constant). This negative effect can partly be offset by decreasing the fixed cost requirement by scaling down the number of product varieties. The other comparative statics results need no further discussion.

The above results make clear that the social planner faces a trade-off between, on the one hand, large variety with positive productivity effects in the assembly sector and, on the other hand, high growth rates in the intermediate goods sector. The greater the returns to variety (the larger σ), the more growth the social planner will sacrifice in exchange for variety. It is at this point that our model differs crucially from Grossman and Helpman (1991; Chapter 3). In their model, growth is growth in product variety. Large returns to variety thus result in sub-optimally low rates of growth (see Benassy, 1998 and De Groot and Nahuis, 1998).

4.2 Market equilibrium versus the social optimum

We are now ready to compare the market outcome as discussed in Section 3 with the FBSO. The model is characterized by three potentially distorting market failures. First, there is a (static) distortion in allocation resulting from the market power that intermediate goods producers have, leading them to engage in mark-up pricing. The second distortion results from the fact that entrants in the intermediate goods sector ignore the productivity effect on the producers of consumer goods (the diversity effect). And third, entrants do not take into account that they decrease the effective market size for their competitors (the business stealing effect). Figure 7.1 depicts the comparison between the market equilibrium and the social optimum. Panel (A) depicts the growth rate in the market equilibrium (g^M) and in the social optimum (g^{FB}), whereas Panel (B) depicts the number of firms (product diversity) in the market equilibrium (N^M) and in the social optimum (N^{FB}). The figure is constructed using equations (7.19), (7.20), and (7.28).[15]

With respect to the rate of growth, we conclude that the market rate of growth is optimal at $\sigma=\varepsilon/(\varepsilon-1)$. This is the specific value for the preference for diversity implicitly assumed by Dixit and Stiglitz (1977). The number of product varieties supplied by the market (N^M) is sub-optimally low at this point, which basically restates the Dixit-Stiglitz result. Market power of the intermediate goods producers results in non-competitive pricing, whereas labour can be hired in a perfectly competitive labour market. Therefore, the use of inputs is distorted in the direction of labour in the assembly sector (L_C/NL_x is too high from a social point of view). Aggregate demand for intermediates is too and this results in too little variety of intermediates. The growth rate is socially optimal in the market economy, as in this special case the elasticity of demand reflects *both* excess entry due to business stealing *and* the contribution of variety to overall productivity.[16] Hence, the product diversity effect and the business stealing effect happen to be of equal magnitude (cf. Grossman and Helpman (1991, Appendix A.3.3)).

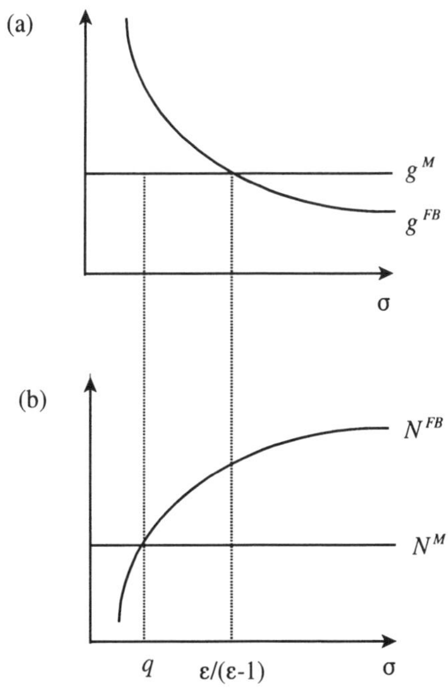

Figure 7.1 *The market equilibrium versus the social optimum*

To the right of the intersection of g^M and g^{FB}, the increasing returns to variety are strong but *not* (fully) reflected in the market by a high mark-up (as in the special case discussed before). The relatively low mark-up implies that profit opportunities are gloomy and hence that the market supports only a low number of firms (see panel (B) of Figure 7.1). Thus, the number of firms in the market is sub-optimally low. Investing in R&D is a fixed cost that can be spread over more sales in case the market is larger. Hence, the existence of a low number of firms makes the return to R&D high and causes the market to produce a sub-optimally high growth rate. To the left of the intersection of the curves N^M and N^{FB}, the argument presented above can be reversed; when the returns to variety are low, the market provides too much variety and growth is too low. In the intermediate case where $q<\sigma<\varepsilon/(\varepsilon-1)$, the market provides too little variety and too low growth. The intuition for this result can be explained in three steps. (1) In the parameter range under consideration, the social return to variety is relatively low. The entry decision, however, is based on a price elasticity (ε) that is relatively low (profit opportunities are relatively high). This suggests a tendency for excessive entry. (2) A countervailing power arises through the distortionary pricing by differentiated goods producers. Cost minimization of the assembly firm then leads to too low aggregate demand for intermediates. This biases the entry decision downward. On balance entry is sub-optimally low. (3) As we have seen, non-competitive pricing as such does not impede a Pareto optimal growth rate (see panel (A) at $\sigma=\varepsilon/(\varepsilon-1)$). For the growth rate we only have to take into account the part of the reasoning under (1).[17] This implies, following the logic expressed above, a sub-optimally low rate of growth.

4.3 Decentralizing the first-best social optimum

Having determined the first-best social optimum, we are now ready to consider how a policy that can support the first-best social optimum as a competitive equilibrium should look. Having three distortions in our model, three instruments are in principle required to decentralize the FBSO. However, as we will show, in this specific model two instruments suffice.[18] We introduce two policy instruments, namely an output subsidy on intermediates (s) and a lump sum tax levied on firms producing the intermediates (T).[19] We assume that the revenues that the government needs in order to be able to pay the subsidies are generated in a non-distortionary lump-sum fashion by taxing consumers. Instantaneous profits of firms producing intermediates now equal $\pi = xp_x(1+s)-(L_x+L_r+L_f)w-T$. Following the same procedure as in Sections 2 and 3, we can solve for the equilibrium growth and interest rates, the equilibrium number of intermediate goods producers, and the equilibrium allocation of labour over the various activities (see Appendix C). Putting the equilibrium values equal to those in the first-best social optimum yields the optimal output subsidy and lump-sum tax. As shown in Appendix C, this gives rise to

$$s = \frac{1}{\varepsilon - 1} > 0 \quad \text{and} \quad T = \left[\frac{\varepsilon}{\varepsilon - 1} - \sigma\right] \frac{\xi L_f \beta(\rho - 1) - \theta}{\xi[\beta(\sigma - 1)(\rho - 1) - 1]}. \qquad (7.30)$$

Decentralization of the first-best social optimum thus requires an output subsidy that is larger the more market power producers of intermediates have (that is, the less substitutable the intermediates are). This subsidy corrects for the distorted price of the intermediates due to monopolistic pricing and the associated too low demand for intermediates that results from this price distortion. The lump sum tax (or subsidy, depending on whether the returns to variety as measured by σ is larger or smaller than the mark-up) is required to make firms of the correct size or, to put it alternatively, to sustain the first-best number of firms. It thereby corrects for the profit destruction effect and the product diversity effect. Both the size and sign of the lump sum tax crucially depend on the returns to variety and the elasticity of substitution between intermediates. If returns to variety are relatively large compared to the mark-up ($\sigma > \varepsilon/(\varepsilon\text{-}1)$), a subsidy should be granted to the producers of intermediates ($T<0$), in order to keep more firms alive than could otherwise be sustained. This follows from the fact that excess entry resulting in profit destruction is limited given the limited market power firms have, while the problem of insufficient entry resulting from the returns to variety is large. Hence, there is insufficient entry and a subsidy is required. In case returns to variety are relatively small, a lump-sum tax should be levied. In the special case where the returns to diversity are exactly equal to the mark-up, no lump-sum transfer is needed and the first-best social optimum can be achieved by just granting the producers of intermediates an output subsidy.

To conclude this section, we have seen that the market growth rate is optimal in the special case when $\sigma = \varepsilon/(\varepsilon\text{-}1)$, while the number of varieties served by the market is sub-optimally low. This last result reiterates the Dixit-Stiglitz result of insufficient entry, but now in a dynamic context. In this special case, the FBSO can be replicated in a decentralized economy by granting intermediate goods producers an output subsidy. If the diversity effect is stronger, the market provides a sub-optimally high growth rate and features insufficient entry. Besides an output subsidy, a lump sum subsidy to intermediate goods producers is needed to replicate the FBSO. If, however, the diversity effect is less pronounced, growth is sub-optimally low and the market might support a number of varieties that is either too high or too low from a social viewpoint. Decentralization of the first-best social optimum requires an output subsidy and a lump-sum tax on intermediate goods producers.

5. THE OPTIMAL NUMBER OF PERMITS

Successful implementation of industrial policies to achieve the first-best results derived in the previous section may be difficult and unrealistic. They require enormous amounts of information for the government (and the government to behave in a benevolent manner). In this section, we consider the second-best problem of choosing the welfare maximizing number of firms, taking as given the non-competitive behaviour after entry of intermediate goods producers (this closely resembles the analysis in Mankiw and Whinston (1986), but in a dynamic context). Hence, it is assumed that the social planner cannot control the behaviour of a given number of firms, but can choose the number of allowances. In order to derive the socially optimal number of permits to be put on the market, we derive an explicit expression for welfare (given the behaviour of firms).

Recall the utility function introduced in Section 2

$$U_0 = \int_0^\infty \frac{C_t^{1-\rho}-1}{1-\rho} e^{-\theta t} dt. \tag{7.31}$$

Using

$$C_t = X_t^{\beta} L_{Ct}^{1-\beta}, \tag{7.32}$$

and the fact that the allocation of labour and the growth rate are constant over time, we arrive at

$$U_0 = \int_0^{\infty} \frac{\left[(N^{\sigma} f_0 L_x)^{\beta} L_C^{1-\beta}\right]^{1-\rho}}{1-\rho} e^{[(1-\rho)\beta g - \theta]t} dt, \tag{7.33}$$

where f_0 is the initial productivity level (which subsequently grows at the constant rate g). Integrating this expression finally yields

$$U_0 = \frac{-1}{(\rho - 1)\left[\beta g(\rho - 1) + \theta\right]\left[(N^{\sigma} f_0 L_x)^{\beta} L_C^{1-\beta}\right]^{\rho-1}}. \tag{7.34}$$

The next step in our analysis is to characterize the effect of a change in the number of product varieties on utility. The sign of the derivative of the present discounted utility with respect to N equals

$$\begin{aligned} &\text{sgn.}\left(\frac{dU_0}{dN}\right) = \\ &\text{sgn.}\left(\sigma\beta\frac{1}{N} + \beta\frac{1}{L_x}\frac{\partial L_x}{\partial N} + (1-\beta)\frac{1}{L_C}\frac{\partial L_C}{\partial N} + \frac{\beta}{\beta g(\rho-1)+\theta}\frac{\partial g}{\partial N}\right). \end{aligned} \tag{7.35}$$

Utility is affected by an increase in the number of varieties of the intermediates (that is, the number of permits issued by the government) through four channels[20]:

1. There is a direct *positive* productivity effect (the first term); this is due to the diversity effect in the Ethier specification. This effect is increasing in

σ, the returns to diversity parameter, and decreasing in N, the level of variety already attained.
2. There is a *negative* volume effect (second term) related to business stealing or profit destruction; more varieties reduce the size of firms producing intermediates and thereby reduce the produced volume of intermediates.
3. There is a *negative* effect on consumption goods production (third term); more varieties are resource consuming by the increase of the total fixed cost in the economy, reducing labour available for production of the consumption good in the assembly sector.
4. There is a *negative* growth effect (the fourth term); more varieties reduce the scale of operations for each firm and hence the profitability of engaging in R&D. The weight for this (negative) effect depends negatively on the pure rate of time preference (θ), and positively on the intertemporal elasticity of substitution ($1/\rho$).

The negative effects are essentially caused by two factors: an effect running via the 'potential of the economy to have large intermediate goods producers with a high growth potential' (L/N); and an effect running via the effective supply of labour ($L\text{-}NL_f$) that is affected by increases in fixed costs following an increase in variety.

The optimal number of firms in the second-best social optimum (SBSO) can now be derived as

$$N^{SB} = \frac{\beta(\sigma-1)(\rho-1)-1}{\beta(\rho-1)\xi L_f-\theta}\,\frac{\beta\xi L}{\beta(\sigma-1)+1}. \tag{7.36}$$

Comparing this solution with the solution for the FBSO (equation (7.28)) reveals that the product variety chosen by the planner who has only one instrument at his disposal is exactly equal to the FBSO. This reflects the fact that the marginal utility of an additional variety is independent of the rate of growth. So comparing the SBSO for the number of intermediates with the solution in the decentralized equilibrium gives rise to the same conclusions as described in Section 4.2.

In the special case where $\beta=1$ (that is, only intermediates are used in the assembly sector and hence the potentially distorting market failure due to mark-up pricing becomes ineffective as relative prices in the assembly sector are no longer distorted), the second-best growth rate equals the first best.[21] So when the distortion in factor allocation resulting from mark-up pricing by intermediate goods producers is 'eliminated', the social planner who has only one instrument at his disposal (the number of permits) can perfectly replicate the FBSO.

In the more general case with $\beta<1$, the growth rate in the second best is always lower than in first best (see Appendix D). Finally, comparing the growth rate in the SBSO with the market equilibrium basically yields the same qualitative conclusions as in Section 4.2. In the special case of $\sigma=\varepsilon/(\varepsilon-1)$, the implicitly chosen growth rate by the planner is lower than the market growth rate (remind that the planner chooses the first-best number of firms, which is higher than the market supports). This reflects the predetermined weight on the lack of variety in the Dixit-Stiglitz specification, for which the planner solves.

Figure 7.2 compares all three growth trajectories with the accompanying variety for different values of σ.[22] The comparison of the SBSO with the FBSO shows, as stated above, that the optimal number of product varieties is equal in both cases. As the available instruments are limited in the SBSO it is obvious that, given the equality of the optimal variety, the growth rate always must be lower in the SBSO than in the FBSO. Taken that the optimal number of firms is equal in the two welfare exercises and that a the trade-off between growth and variety exists, the comparison between the market rate of growth and the growth rate attained by SBSO policy is obvious. It should be noted that with a permits policy it is impossible to improve both on the number of varieties and the growth rate at the same time.

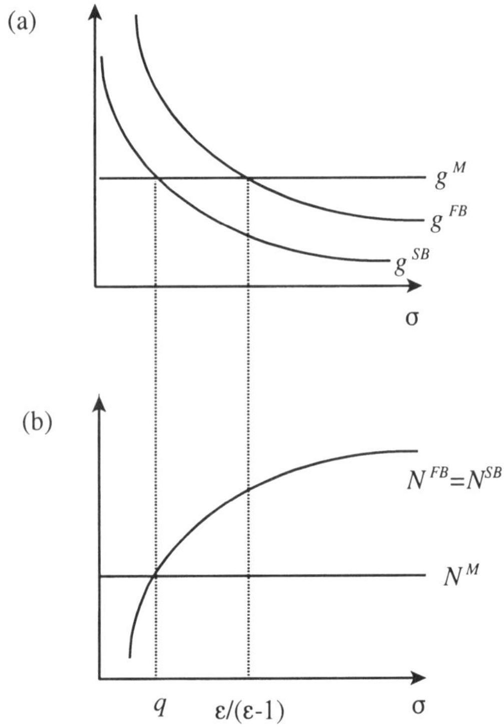

Figure 7.2 *FBSO vs SBSO*

6. KNOWLEDGE SPILLOVERS AND WELFARE

Spillovers of knowledge are persuasively found in empirical research. A spillover is a classical example of an externality, and hence from the point of view of welfare analysis of crucial importance. Empirical research shows that knowledge creation in the form of a granted patent, or research activity in the form of R&D spending of one agent is beneficial for other agents in the economy. This simple argument suggests that R&D activity is too low from a social point of view. The findings in the previous sections rely on a trade-off between growth and variety. This trade-off is a desirable feature of our model that partly emerges from the specific nature of knowledge that we assume. In our analysis so far, knowledge is specific to a firm or organization and in that sense it is excludable. The very nature of knowledge, however, is that it is non-rival (see, for example, Romer (1990), this is extensively discussed in Chapter 1). Given the convincing empirical evidence on the existence of knowledge spillovers we relax the assumption of perfect excludability of knowledge, applied in this chapter so far.

6.1 Modelling knowledge spillovers

To introduce spillovers in the model we postulate the following engine of growth, replacing equation (7.10)

$$\dot{f}_i = \xi f_i^{1-\alpha} Z_i^{\alpha} L_{ri}, \quad Z_i = Z(f_j), \quad j \neq i , \tag{7.37}$$

where 1-α measures the degree to which the returns to investments in R&D are appropriable, and Z_i is the relevant knowledge pool available for firm *i*, which is not accumulated by firm *i* itself. The larger α, the larger the disincentive effect for firms investing in R&D will be. Setting α equal to zero yields the original equation for knowledge accumulation in which the returns to investments in knowledge are fully appropriable. Some features of this functional form are worth noticing. First, and most important for the welfare analysis, is that an externality is introduced. By means of R&D activity all firms contribute to the *common* knowledge pool, a consequence which they do not take into account when determining their behaviour (purposely spilling knowledge is excluded; thus, we abstain from strategic interaction since we assume the number of firms to be large throughout the analysis). Secondly, firm *i* does not benefit from firm *j*'s R&D activity unless it engages in R&D activity itself.[23] Thirdly, if Z_i is time-invariant, R&D activities run into diminishing returns, unless α=0.[24]

The solution procedure of the model is similar to that above, as in the derivation of the first order conditions Z_i is taken parametrically by firms. To get more insight we have to put more structure to the model by being explicit about the knowledge pool. In empirical work different R&D or knowledge stocks are weighted according to 'technological distance' or input-output relations; in our stylized model these concepts are not defined explicitly. We propose

$$Z_i = \sum_{j=1}^{N} N^{\mu-1} f_j, \quad j \neq i, \quad \mu > 0. \tag{7.38}$$

Hence, only the degree of duplication ($\mu<1$) or inspiration ($\mu>1$) is taken into account. If $\mu<1$, the marginal effect of the entrance of other firms with an average knowledge stock is declining. Hence, in terms of the empirical literature, the 'technological distance' of a new firm exceeds the average 'technological distance' in the economy. A different way of understanding what $\mu<1$ means is to think of duplication of already developed knowledge. The 'inspiration' case in which $\mu>1$ implies that contact with an additional firm in possession of the average knowledge stock increases the knowledge pool more than proportionally. Ideas can thus be created through possible 'neue combinationen'.

Assuming symmetry of firms in f_j ($f_j=f$ for all j) and N large, the knowledge accumulation function boils down to

$$\dot{f}_i = \xi N^{\mu\alpha} f_i^{1-\alpha} \bar{f}^{\alpha}, \tag{7.39}$$

where $\bar{f}$ is the average knowledge stock.

6.2 The market solution

Replacing the 'old' engine of growth with equation (7.39), we can solve for the market equilibrium of the model where we proceed along the same lines as before.

Combining the FOC's for firm behaviour with the Ramsey rule and the zero profit condition results in a relation between the rate of growth and the number of firms that looks like

$$g^{ZP} = \frac{(\varepsilon-1)\xi L_f N^{\mu\alpha} - \theta}{\beta(\rho-1) + \alpha - (\varepsilon-1)} . \tag{7.40}$$

This result reveals that the rate of growth that can be obtained is increasing in the number of firms, since each individual firm adds to the knowledge base from which all individual firms in the economy can benefit (unless μ=0 and/or α=0). Combining goods- and labour-market equilibrium, the zero-profit condition and the Ramsey rule, we obtain a second relation between the growth rate and the number of firms that equals

$$g^{LME} = \frac{\dfrac{\xi\beta(\varepsilon-1)L}{\varepsilon N^{1-\mu\alpha}} - \theta}{\beta(\rho-1) + \alpha} . \tag{7.41}$$

Given the operation of the labour market constraint, more firms leave less labour for R&D which will depress the rate of growth, unless the degree of inspiration is large (μ>1/α). Confronting these relations in a graph yields the solution of the model as the intersection of the two curves. The graph is depicted in Figure 7.3.

We will now consider three cases. In the first case, μ equals zero and α is equal to zero. This is the case considered in Section 3. In this case, g^{ZP} is independent of the number of firms. In terms of Figure 7.3, g^{ZP} is horizontal. Though an increase in the size of the economy increases the number of firms (g^{LME} shifts outward in Figure 7.3), the growth rate remains unaffected. In the second case, we leave μ equal to zero, but we allow for knowledge spillovers by allowing α to be larger than zero.[25]

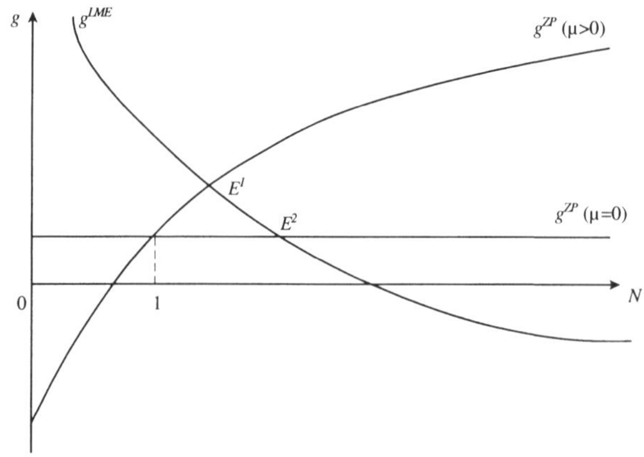

Figure 7.3 General equilibrium with spillovers

The rate of growth and the equilibrium number of firms in this case can be derived as

$$g = \frac{(\varepsilon - 1)\xi L_f - \theta}{\beta(\rho - 1) + \alpha - (\varepsilon - 1)}, \quad N = \frac{\xi\beta L[\beta(\rho - 1) + \alpha - (\varepsilon - 1)]}{\varepsilon[\xi L_f(\beta(\rho - 1) + \alpha) - \theta]}. \tag{7.42}$$

An increase in α, that is spillovers become more prominent, will decrease the rate of growth and increase the number of firms.[26] Less R&D activity means a lower fixed cost burden resulting from R&D expenses (further denoted as a 'quasi-fixed' cost), and hence entry is higher with spillovers than without spillovers. The model is still characterized by the absence of scale effects since g^{ZP} is still horizontal, and the trade-off between growth and variety prevails. Finally, we consider the case in which μ is also taken to be positive. In this case, g^{ZP} is upward sloping and the rate of growth is no longer independent of the scale of the economy. Due to non-linearities, a closed form solution for the rate of growth and the equilibrium number of firms cannot be obtained.

To get some feeling for the forces that are at play when the number of firms change, we will now solve the model taking the number of firms exogenous. Starting from the first order conditions and the Ramsey rule and using labour market equilibrium the equilibrium growth rate is found as[27]

$$g = \frac{\xi N^{\mu\alpha}\left[\frac{L}{N} - L_f\right] - \theta\frac{\varepsilon - \beta}{\beta(\varepsilon - 1)}}{1 + \left[\beta(\rho - 1) + \alpha\right]\frac{\varepsilon - \beta}{\beta(\varepsilon - 1)}} \tag{7.43}$$

Taking the derivative of the growth rate w.r.t. the number of firms, it follows that the growth rate is a decreasing function of N, unless $\mu\alpha[L/N-L_f]>L/N$, that is unless the degree of inspiration is very large, the disincentive effect is large and fixed costs are low. In this special case, the trade-off between growth and variety no longer prevails.

6.3 Market equilibrium compared to the first-best social optimum with spillovers

An analytical comparison of the market equilibrium with the first-best social optimum for the general case is intractable due to non-linearities in the expressions for the growth rate, the allocation of labour and the equilibrium number of firms. Appendix E provides some computational details on the implicit functions describing the first-best and market equilibrium. In the remainder, we restrict attention to the main conclusions and focus on the intuition behind the results we obtain. In order to focus on the externality resulting from spillovers, we now exclude distortionary pricing (by setting β equal to one), and we restrict attention to the Dixit-Stiglitz specification where the productivity effect and the business stealing effect cancel (that is $\sigma=\varepsilon/(\varepsilon-1)$). The knowledge spillover is thus the only remaining distortion in the economy.[28]

Looking at the disincentive effect in isolation ($\alpha>0$ and $\mu=0$), growth is sub-optimally low and variety is sub-optimally high. Starting off here, a ceteris paribus increase of μ results in an increase in the growth rate, as N

enters the knowledge base. It can be shown that for some $\mu>1$ the market rate of growth is too high from a social point of view and that variety is too low, the opposite result of the case where the disincentive effect operates alone. The logic here is simple. With $\mu = 0$ growth is lower than optimal due to the disincentive effect and consequently too many firms can survive in the market. At some μ, increases in variety are so potent in the growth engine – the inspiration externality of new entrants – that lower input of R&D in exchange for more firms is to be preferred from a social point of view.[29]

Figure 7.4 provides a welfare analysis in the $(1-\alpha)$-μ-space. The figure is based on numerical analysis (sensitivity analysis has revealed that the qualitative results are insensitive for changes in parameters). Details and parameter values are provided in Appendix E. Every point in the grid represents a different $(1-\alpha)$-μ combination keeping the other parameters fixed. At every point the growth rate and variety supported by the market are compared to the optimal growth rate and optimal variety. The lines in the figure demark regime switches with respect to optimality.

We initially focus on the vertical dimension in the figure: starting at a certain value for $1-\alpha$, we increase μ from zero. At $\mu=0$ only the disincentive effect is present and the effort put into research is too low and hence the 'quasi-fixed' research cost is low and entry is sub-optimally high. At the NN-line the number of products sustained in the market is optimal. Above the NN-line, μ is that high, implying variety is so potent in the growth engine, that despite the disincentive effect entry is too low from a social point of view. Remember that growth is too low due to the disincentive effect. This as such implies a relatively low fixed cost. But variety is so powerful in the knowledge base that the social planner prefers to have more firms in the market. Increasing μ even further, above the GG-line where the market rate of growth equals the optimal rate of growth, takes the economy into a third regime.

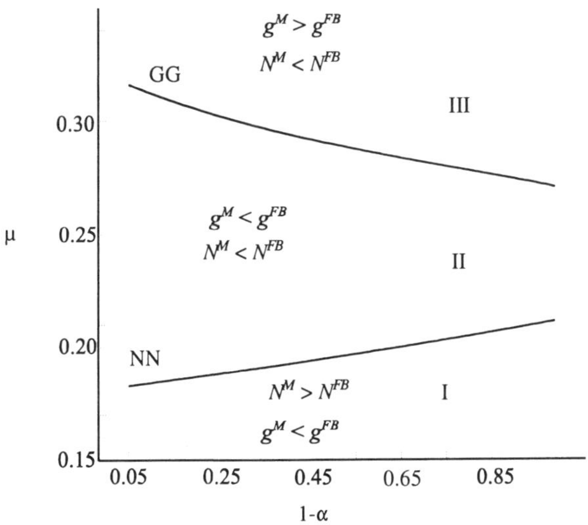

Figure 7.4 *Welfare and spillovers*

In this regime the market rate of growth is too high and the supported variety remains sub-optimally low. In this regime the planner would decrease research effort, as less effort implies less 'quasi-fixed' costs and saved resources. Again, the decrease in the 'quasi-fixed' cost induces entry. From the planner's perspective it allows for more variety. And note that variety is, in this parameter range, extremely potent in the research engine. This counteracts the welfare loss due to the ceteris paribus decrease in the growth rate. In sum, if the elasticity of variety with respect to the growth rate is extremely high, the planner is willing to sacrifice growth for increased variety, which makes the market growth rate too high despite the disincentive effect.

Taking a horizontal perspective might add some further insight. The upward slope of the NN-curve can be explained as follows. Starting off from a point on the NN-curve where the number of firms in the market is at its optimum, a decrease in α has two effects. First, it lessens the disincentive effect, implying that firms in the market increase their R&D which increases their fixed costs and thereby results in too little entry in the market compared to the social optimum. This worsens the problem of insufficient entry in the market equilibrium. Secondly, it lessens the problem of insufficient entry in the market since $\mu\alpha$ and thus the degree of inspiration has decreased. Since an increase in μ is required to arrive in a new situation where the number of firms in the market is again in its social optimum, we conclude that the latter effect of α is dominant. The upward slope of the NN-curve is confirmed by extensive numerical experimentation with the model for reasonable parameter values.

Along similar lines, we can explain the downward slope of the GG curve. Starting at a point on this curve, a decrease in α has two effects. Firstly, it reduces the disincentive effect and thus makes growth too large from a social point of view. Secondly, it reduces $\mu\alpha$ which means that entry becomes sub-optimally large, firms become sub-optimally small and growth becomes sub-optimally small. Since a decrease in μ is required to bring growth back to its social optimum, we may conclude that the former effect is dominant.

This section shows that, unless the degree of 'inspiration' is very large, the results derived in our analysis in the previous sections stand upright.

7. CONCLUSION

This chapter has developed a simple two-sector model of endogenous growth. Increased variety of intermediates was shown to have both a positive effect on welfare via increased productivity and negative effects via reduced growth and reduced availability of productive labour. Three potential market failures where distinguished, that is, a product diversity effect, a business stealing effect, and a pricing distortion. The market was shown not to provide in any case an equilibrium that reflects a socially desirable trade-off between variety and growth. When returns to diversity are strong relative to the toughness of competition, the market supports too high a growth rate and yields a lack of variety. When the returns to diversity are very weak this conclusion is reversed. The intermediate case might deliver a growth rate and a number of varieties that are both too low from a social point of view. These results illustrate the importance of disentangling the returns to variety and the imperfect substitutability of brands when analysing the welfare properties of models of monopolistic competition with endogenous growth. We believe that a trade-off between growth and variety is of crucial importance with respect to optimal R&D policy. This trade-off that is at the heart of our previous analysis and is robust to the introduction of knowledge spillovers. Introducing spillovers makes the rather uncommon finding – that variety is too low from a social point of view and growth too high – even more likely. Only in the 'inspiration' regime – a very high degree of positive interaction between knowledge developed by different firms in the economy is present – the trade-off between growth and variety vanishes.

NOTES

1. We have no pretension of being exhaustive in the overview to follow.
2. A second difference is, as discussed earlier, that we follow Ethier (1982) in assuming that there are returns to variety in the production of final goods instead of in consumption goods, as is the case in Dixit and Stiglitz (1977), and van de Klundert and Smulders (1997).
3. The notion of business stealing has strong resemblance to the *profit destruction effect* that Grossman and Helpman (1991) distinguish.
4. In an informal discussion on variety and growth, Young (1998) hints at this result in case one deviates from Dixit-Stiglitz preferences.
5. Yang and Heijdra (1993) criticize this approach. They argue that the result that the number of firms is so large that competition is monopolistically à la Chamberlin should be the *outcome* of the model and not be imposed a priori. To avoid this problem, we would have to introduce the concept of a perceived price elasticity, where this perceived elasticity depends on the number of competitors (see Smulders and van de Klundert (1995)). As this extension would seriously complicate the analysis and not affect the main conclusions of the chapter, we abstain from this issue.
6. The process of entry and exit is assumed to drive profits to zero in our model. This can be extended by explicitly introducing endogenous entry costs, see Peretto and Smulders (2002). In our analysis we abstain from explicitly modelling entry costs; our results are insensitive for an extension of the model in such a direction.
7. Subscript t denotes time.

8. The assets W consist out of consumer loans and shares issued by high-tech firms to finance their investments in research. As in a consolidated equilibrium, net debt among consumers is zero, the income on the assets, rW, equals the dividends paid by high-tech firms. We will later return to the savings-investment equilibrium when discussing the general equilibrium of the model.
9. It is easily demonstrated that in case $\sigma=\varepsilon/(\varepsilon-1)$, the expression for the composite good boils down to:

$$X = \left[\sum_{i=1}^{N} x_i^{\frac{\varepsilon-1}{\varepsilon}}\right]^{\frac{\varepsilon}{\varepsilon-1}}, \text{ so in symmetry } X = N^{\frac{1}{\varepsilon-1}}(Nx).$$

This specification is used in many subsequent analyses (e.g., Grossman and Helpman (1991) and van de Klundert and Smulders (1997)). We prefer the more general Ethier specification as we see no need for a strict one to one relation between the mark-up resulting from market power and the returns to diversity. See Benassy (1996) for a similar argument.
10. Stability of the equilibrium with a positive growth rate requires $\beta(\rho-1)>(\varepsilon-1)>\theta/(\xi L_f)$.
11. Labour productivity in the assembly sector (C/L_c) *grows* at rate βg. The labour productivity *level* equals:

$$\frac{C}{L_C} = \left(N^{\sigma-1} f \frac{\beta}{1-\beta}\frac{(\varepsilon-1)}{\varepsilon}\right)^{\beta},$$

which positively depends on N (and thus on L) as $\sigma>1$. This result basically reflects Adam Smith's notion of division of labour. The result that there are no dynamic returns to scale in this model depends on the specific choice of our knowledge base (equation (7.10)). A more extended knowledge base in which the equilibrium number of firms features, alters this result. See Section 6.
12. The model allows for a $\sigma\geq 1$ restriction, but with equality the social planners' problem is not well defined. So here the parameter restriction needs to be narrowed to $\sigma>1$. It is intuitively clear that it is socially optimal to have a single variety in case $\sigma=1$. As knowledge is completely firm specific and there is no return to variety as such, it is optimal to minimize on the total fixed cost in the economy. To allow for $\sigma=1$, the restriction $N \geq 1$ should be added to the social planners' optimization programme.
13. Use that under symmetry $X = N^{\sigma-1}Nx = N^{\sigma}fL_x$.
14. We impose the parameter restriction $1/(\beta(\rho-1))<\sigma-1<\xi L_f/\theta$. Derivation of the optimal number of intermediate goods producers can be found in Appendix B.
15. From equations (7.19), (7.20), and (7.28), we can derive the partial derivatives of the growth rate and number of firms with respect to σ: $\partial g^M/\partial\sigma=\partial N^M/\partial\sigma=0$, $\partial g^{FB}/\partial\sigma<0$, $\partial^2 g^{FB}/\partial\sigma^2>0$, and $\partial N^{FB}/\partial\sigma>0$, $\partial^2 N^{FB}/\partial\sigma^2<0$. The point of intersection of g^{FB} and g^M is at $\sigma=\varepsilon/(\varepsilon-1)$.The point of intersection of N^{FB} and N^M (point q in Figure 7.1) is at $\sigma=1+[\beta(\rho-1)+1]/[\beta\{\rho(\varepsilon-\beta)-(1-\beta)\}]$. By straightforward calculation it follows that $1+1/(\beta(\rho-1))<q<\varepsilon/(\varepsilon-$

1).

16. Another way of putting this is to say that firm *size* is optimal, and the decision to invest in R&D is not distorted in any other way. There is no appropriability problem as there are no spillovers, hence the firm's knowledge base is independent of the number of firms. The private surplus from innovation equals the social surplus.
17. This reasoning can be illustrated as follows: if mark-up pricing (leading to too low aggregate demand for intermediates) would evaporate, the distortions on growth and on variety would be balanced again. This is easily demonstrated by setting β equal to one (i.e., no direct labour is used in the production of consumption goods). It then holds that $q=\varepsilon/(\varepsilon-1)$, so the area between q and $\varepsilon/(\varepsilon-1)$ in Figure 7.1 vanishes. Note the similarity of this version of the model (with $\beta=1$ and $\sigma=\varepsilon/(\varepsilon-1)$) with the model of Grossman and Helpman (1991, Chapter 3).
18. This result resembles the result in Barro and Sala-i-Martin (1995, Chapter 6) where in a more or less similar model to ours (without explicitly distinguishing between returns to variety and the elasticity of substitution) a static and dynamic inefficiency can be eliminated with a single instrument. A similar issue arises in Grossman and Helpman (1991, Chapter 3) where three distortions can be eliminated using one instrument.
19. We could as well introduce a subsidy on the use or purchasing of intermediates in the assembly sector. It is easily verified that this subsidy should equal $1/\varepsilon$. This subsidy is required to correct for the distorted price of the intermediates due to monopolistic price setting.
20. Note that in performing this analysis, we look at an *exogenous* change in N. The number of firms is determined exogenously. We assume in other words that entry is blocked (or, alternatively, exit not required). The solution of the model under the assumption that there is no free entry is given in Appendix D.
21. We refer to Appendix D for the solution of the growth rate.
22. The curve for the SBSO growth rate is drawn using the facts that $\partial g^{SB}/\partial\sigma<0$ and $\partial^2 g^{SB}/\partial\sigma^2>0$. We refer to Appendix D for a derivation.
23. Cohen and Levinthal (1989) stress the importance of R&D to learn to absorb knowledge developed by other agents in the economy.
24. This fundamental change of putting $1-\alpha$ smaller than 1 is introduced as in the symmetric equilibrium Z_i grows at the same rate as f_i, and hence to analyse a steady state with spillovers and a constant growth rate we need to introduce $(1-\alpha)<1$.
25. This case is examined in van de Klundert and Smulders (1997).
26. The effect on the number of firms follows from explicitly solving for N and taking the derivative with respect to α. This reveals that the equilibrium number of firms unambiguously declines in α.
27. Use that $L=N(L_x+L_r+L_f)+(1-\beta)\varepsilon/\beta(\varepsilon-1)NL_x$ where L_x and L_r satisfy equation (E.4) and $r=\beta(\rho-1)g+\theta$.

28. In terms of Figure 7.1 q equals $\varepsilon/(\varepsilon\text{-}1)$ so the region between q and $\varepsilon/(\varepsilon\text{-}1)$ disappears and we focus only on the vertical line through $\varepsilon/(\varepsilon\text{-}1)$ on the horizontal axis. Hence both growth and variety are, but for knowledge spillovers, optimally supplied by the market.
29. The market thus creates a fixed cost burden that is too high for social optimal entry.

APPENDICES

A. Solution of the model

In this appendix, we characterize the steady-state solution of the model. Using the expressions for the price of intermediates (equation (7.13)) and the allocation rule of research labour (equation (7.14)), we can derive

$$\frac{\dot{p}_x}{p_x} = \frac{\dot{w}}{w} - \frac{\dot{f}}{f} = \frac{\dot{w}}{w} - g = \frac{\dot{q}_f}{q_f}. \tag{A.1}$$

Note that the wage rate is the numéraire (w=1). In the steady state, the allocation of labour is constant, and the number of firms is fixed. This gives rise to

$$g_C = \beta g_x = \beta g \quad \text{and} \quad \frac{\dot{P}_C}{P_C} = \beta \frac{\dot{p}_x}{p_x} = -\beta g. \tag{A.2}$$

So we can write the Ramsey rule as

$$\beta g = \frac{r + \beta g - \theta}{\rho}. \tag{A.3}$$

This equation corresponds to equation (7.18) in the main text.

Using equations (7.13), (7.14), and (A.1), the no-arbitrage condition (7.15) yields

$$-g - r = -\xi(L_x + L_r). \tag{A.4}$$

Imposing free entry and exit in the intermediate goods sector results in zero excess profits (equation (7.17)). Substituting the price for intermediates (equation (7.13)) and equation (7.9) into the zero profit condition (equation

(7.17)), it boils down to

$$\frac{L_x + L_r + L_f}{L_x} = \frac{\varepsilon}{\varepsilon - 1}. \tag{A.5}$$

The firm size in relation to the size of the production department is thus equal to the mark-up. The larger the mark-up, the more a firm can afford to have large fixed costs (L_r+L_f) without making losses. The rate of return on investment can now be written as[1]

$$r = (\varepsilon - 1)(g + \xi L_f). \tag{A.6}$$

This corresponds to equation (7.18) in the main text. Confronting the realized and required rates of returns yields the equilibrium growth and interest rate (equation (7.19) in the main text). The equilibrium number of high-tech firms and the allocation of labour can now be determined using labour market equilibrium (equation (7.16)), and

$$CP_C = \frac{Np_{xi}x_i}{\beta} = \frac{L_C w}{1-\beta} \quad \text{so} \quad \frac{NL_x \frac{\varepsilon}{\varepsilon - 1}}{\beta} = \frac{L_C}{1-\beta}, \tag{A.7}$$

which shows how the assembly sector optimally chooses between labour and intermediates. This leaves us with

$$N = \frac{[\beta(\rho - 1) - (\varepsilon - 1)]\xi\beta L}{\varepsilon[\beta(\rho - 1)\xi L_f - \theta]} \quad \text{and} \quad L_C = (1-\beta)L. \tag{A.8}$$

This equation corresponds to equation (7.20) in the main text.

B. The first-best social optimum

This appendix solves for the optimal number of firms in the intermediate goods sector in the first-best social optimum. Differentiation of equation (7.24) with respect to time yields

$$\frac{\dot{q}_f}{q_f} = (1-\rho)\frac{\dot{C}}{C} + \frac{\dot{N}}{N} - \frac{\dot{L}_C}{L_C} - \frac{\dot{f}}{f}. \tag{B.1}$$

We know by combining the equations (7.24) and (7.26) that

$$\frac{\dot{q}_f}{q_f} = \theta - \xi L_r - \frac{\beta \xi L_C}{N(1-\beta)}. \tag{B.2}$$

By differentiation of equation (7.3) with respect to time, we get (using symmetry among the intermediate goods producers)

$$\begin{aligned} \frac{\dot{C}}{C} &= \beta\frac{\dot{X}}{X} + (1-\beta)\frac{\dot{L}_C}{L_C} \\ &= \beta\left[(\sigma-1)\frac{\dot{N}}{N} + \frac{\dot{N}}{N} + \frac{\dot{f}}{f} + \frac{\dot{L}_x}{L_x}\right] + (1-\beta)\frac{\dot{L}_C}{L_C}. \end{aligned} \tag{B.3}$$

We now use the fact that L_C and NL_x are constant to arrive at

$$\frac{\dot{N}}{N}\left[1 + \beta(1-\rho)(\sigma-1)\right] = \theta - \frac{\beta\xi L_C}{N(1-\beta)} - \beta(1-\rho)\frac{\dot{f}}{f}. \tag{B.4}$$

Finally, using the solutions for L_C and the growth rate of f (equations (7.25) and (7.27), respectively), we arrive at the following differential equation in N

$$\dot{N} = \frac{\beta(\rho-1)\xi L_f - \theta}{\beta(\rho-1)(\sigma-1) - 1}N - \frac{\beta\xi L}{\beta(\sigma-1)+1}. \tag{B.5}$$

Under the restriction $1/(\beta(\rho-1))<\sigma-1<\xi L_f\ /\theta$ the root of the differential equation is positive and the optimum is characterized by a positive and constant number of firms. This optimum is obtained by setting $\dot{N}$ equal to zero. This yields equation (7.28) in the main text.

C. Decentralization of the first-best social optimum

In this appendix, we derive a policy mix that can be used to support the first-best outcome as a competitive equilibrium. We introduce an output subsidy on intermediates (s) and a lump-sum tax on firms in the intermediate goods sector (T). The current value Hamiltonian then reads as

$$H = xp_x(1+s) - (L_x + L_r + L_f)w - T + q_f \xi f L_r . \tag{C.1}$$

The first-order conditions can now be determined as

$$\frac{\partial H}{\partial L_x} = f\left(\frac{\varepsilon - 1}{\varepsilon}\right) p_x(1+s) - w = 0 , \tag{C.2}$$

$$\frac{\partial H}{\partial L_r} = -w + q_f \xi f = 0 , \tag{C.3}$$

$$\frac{\partial H}{\partial f} = rq_f - \dot{q}_f = L_x p_x(1+s)\frac{\varepsilon - 1}{\varepsilon} + q_f \xi L_r . \tag{C.4}$$

The solution procedure is now the same as in Appendix A. Using the zero-profit condition, the Ramsey rule (equation (A.3)) and labour market equilibrium (equation (7.16)), we derive the rate of growth, the interest rate, the number of workers in each intermediate goods producing firm, and the equilibrium number of firms as

$$g = \frac{\xi(\varepsilon-1)(L_f+T)-\theta}{\beta(\rho-1)-(\varepsilon-1)}, \quad r = \frac{(\varepsilon-1)\left[\beta(\rho-1)\xi(L_f+T)-\theta\right]}{\beta(\rho-1)-(\varepsilon-1)}, \tag{C.5}$$

$$N = \frac{(1+s)\xi\beta L\left[\beta(\rho-1)-(\varepsilon-1)\right]}{(1+\beta s)\varepsilon[\beta\xi L_f(\rho-1)-\theta]+T\xi\beta[(\rho-1)[(\varepsilon-\beta)+\beta s(\varepsilon-1)]+(\varepsilon-1)(1+s)]} \tag{C.6}$$

The solutions for the first-best output subsidy and lump-sum tax are now easily found by setting these solutions for the growth rate and the equilibrium number of firms equal to those in the first-best social optimum (equations (7.27) and (7.28)). The first-best output subsidy and lump-sum tax read as

$$s = \frac{1}{\varepsilon-1} \quad \text{and} \quad T = \left[\frac{\varepsilon}{\varepsilon-1}-\sigma\right]\frac{\xi L_f\beta(\rho-1)-\theta}{\xi\left[\beta(\sigma-1)(\rho-1)-1\right]} \tag{C.7}$$

These are the equations given in the main text.

D. The blocked-entry model (*N* fixed)

This appendix solves the model under the assumption that the number of firms is fixed (equation (A.5) is replaced by $N=\bar{N}$). The system of equations from which the solution can be found consists of equations (A.3), (A.4), (A.7), (7.10) (combined with the definition for the rate of growth), (7.16), and $N=\bar{N}$. We can solve this system of six equations with six unknowns (g, r, L_x, L_r, L_C, N).

The solution for the growth rate is

$$g = \frac{\xi\frac{L}{\bar{N}} - \xi L_f - \frac{\varepsilon-\beta}{\beta(\varepsilon-1)}\theta}{(\rho-1)\frac{\varepsilon-\beta}{\varepsilon-1} + 1}. \tag{D.1}$$

Combining the Ramsey rule and the fact that $L_x = r/\xi$ (see note 1), we can derive

$$L_x = \beta(\rho - 1)\frac{\frac{L}{\bar{N}} - L_f + \frac{1}{\beta(\rho - 1)}\frac{\theta}{\xi}}{(\rho - 1)\frac{\varepsilon - \beta}{\varepsilon - 1} + 1}. \tag{D.2}$$

Finally, we derive labour used in the consumption goods sector as

$$L_C = (1 - \beta)\bar{N}\frac{\varepsilon}{\varepsilon - 1}(\rho - 1)\frac{\frac{L}{\bar{N}} - L_f + \frac{1}{\beta(\rho - 1)}\frac{\theta}{\xi}}{(\rho - 1)\frac{\varepsilon - \beta}{\varepsilon - 1} + 1}. \tag{D.3}$$

Taking derivatives yields

$$\frac{\partial L_x}{\partial \bar{N}} < 0; \quad \frac{\partial L_C}{\partial \bar{N}} < 0; \quad \frac{\partial g}{\partial \bar{N}} < 0. \tag{D.4}$$

The optimum is found by putting $\mathrm{d}U_0/\mathrm{d}N$ equal to zero (see equation (7.35)) which results in

$$N^{SB} = \frac{(\beta(\rho - 1)(\sigma - 1) - 1)L}{[(1 - \beta) + \beta\sigma]\left[L_f(\rho - 1) - \frac{\theta}{\xi\beta}\right]}. \tag{D.5}$$

Substituting this solution for the second-best number of firms into equation (D.1) yields the solution for the second-best rate of growth

$$g^{SB} = (\varepsilon-1)\frac{\xi L_f\left[\dfrac{\rho}{\beta(\rho-1)(\sigma-1)-1}\right]}{(\varepsilon-\beta)(\rho-1)+(\varepsilon-1)} - \frac{\theta\left[\dfrac{\beta\rho(\sigma-1)(\varepsilon-\beta)-\beta(\sigma-1)(1-\beta)-(1-\beta)}{\beta(\varepsilon-1)(\beta(\rho-1)(\sigma-1)-1)}\right]}{(\varepsilon-\beta)(\rho-1)+(\varepsilon-1)}. \tag{D.6}$$

Finally, we compare the growth rate in the second-best social optimum with the first-best rate of growth (equation (7.28)). This yields

$$g^{SB} - g^{FB} = \frac{(1-\beta)\left[\dfrac{\theta}{\beta} - \xi L_f(\rho-1)\right] - 2\theta(\sigma-1)\left[(\rho(\varepsilon-\beta)-(1-\beta)\right]}{\left[\beta(\rho-1)(\sigma-1)-1\right]\left[(\varepsilon-\beta)(\rho-1)+(\varepsilon-1)\right]}. \tag{D.7}$$

Using the conditions $\beta(\rho-1)>(\varepsilon-1)>\theta/(\xi L_f)$ and $\beta(\rho-1)>1/(\varepsilon-1)>\theta/(\xi L_f)$, it is easily seen that the growth rate in the SBSO is always lower than in the FBSO. Equation (D.1) gives the market rate of growth as a function of the number of intermediate goods producers. Using the notion that $N^{SB}=N^{FB}=N^{M}$ at the point q (see note 15), it is evident that $g^{M}=g^{SB}$ at $\sigma=q$. When returns to diversity are relatively strong ($\sigma>q$), growth in the market is larger than growth in the FBSO, whereas the opposite holds if $\sigma<q$. Using $\partial g/\partial N < 0$ (see equation (D.1)) and $\partial N^{SB}/\partial\sigma > 0$, we derive $\partial g^{SB}/\partial\sigma < 0$. Along the same procedure, it follows that $\partial^2 g^{SB}/\partial\sigma^2 > 0$.

E. Analysis of the model with spillovers

The first-order conditions for the market solution with spillovers are obtained performing dynamic optimization:

$$\frac{\partial H}{\partial L_{xi}} = 0 \;\leftrightarrow\; p_{xi} = \frac{\varepsilon}{\varepsilon-1}\frac{w_i}{f_i}, \tag{E.1}$$

according to which firms engage in mark-up pricing,

$$\frac{\partial H}{\partial L_{ri}} = 0 \quad \leftrightarrow \quad w_i = q_{hi}\xi f_i^{1-\alpha}\left(N^{\mu}\bar{f}\right)^{\alpha}, \tag{E.2}$$

showing that firms allocate R&D labour to this sector as long as the marginal benefits of doing so exceed the marginal costs, and

$$\frac{\dot{p}_{hi}}{p_{hi}} + \frac{\partial H}{\partial h_i}\frac{1}{p_{hi}} = \frac{\dot{p}_{hi}}{p_{hi}} + L_{xi}\frac{p_{xi}}{p_{hi}}\frac{\varepsilon - 1}{\varepsilon} + \alpha\xi L_{ri}\left(N^{\eta}\frac{\bar{h}}{h_i}\right)^{1-\alpha} = r, \tag{E.3}$$

which is the no-arbitrage condition. The model is closed by imposing full employment labour market equilibrium and imposing instantaneous profits in the intermediate goods sector (π) to equal zero. Imposing symmetry and using that $\dot{q}_f / q_f = -g$, we can then derive, by combining the first-order conditions of firm behaviour that:

$$L_x = \frac{r + (1-\alpha)g}{\xi N^{\mu\alpha}} \quad \text{and} \quad L_r = \frac{g}{\xi N^{\mu\alpha}} \tag{E.4}$$

Combining these expressions with the Ramsey rule and the zero-profit condition results in relation (7.40).

The solution to the following maximization programme yields the allocation of labour and number of firms that maximizes the utility of the representative agent:

$$H = \frac{1}{1-\rho}\left[\left(L - N(L_x + L_r + L_f)\right)^{1-\beta}\left(N^{\sigma} f L_x\right)^{\beta}\right]^{1-\rho} + q_f \xi f N^{\mu\alpha} L_r. \tag{E.5}$$

The first order conditions corresponding to this problem are

$$\frac{\partial H}{\partial L_x} = 0 \quad \Leftrightarrow \quad (1-\beta)NL_x = \beta L_C, \tag{E.6}$$

$$\dot{q}_f + \frac{\partial H}{\partial f} = \theta q_f \quad \Leftrightarrow \quad \dot{q}_f + \frac{C^{1-\rho}\beta}{f} + q_f \xi N^{\mu\alpha} L_r = \theta q_f. \tag{E.7}$$

$$\frac{\partial H}{\partial L_r} = 0 \quad \Leftrightarrow \quad p_h = (1-\beta)\frac{C^{1-\rho}N}{L_C \xi N^{\mu\alpha} f}, \tag{E.8}$$

$$\frac{\partial H}{\partial N} = 0 \quad \Leftrightarrow \quad C^{1-\rho}\left[\beta\sigma - (1-\beta)\frac{L-L_C}{L_C}\right] + q_f \alpha\mu\xi f L_r N^{\mu\alpha} = 0, \tag{E.9}$$

By substituting equations (E.6) and (E.8) into (E.9), we derive an expression for L_C in terms of L, L_r and N. Substituting this expression and equation (E.6) into the equation for labour market equilibrium, and using the knowledge accumulation equation (7.39) we derive a relation between the growth rate and the number of firms:

$$g_{STAT}^{FB} = \xi N^{\mu\alpha}\left[\frac{\beta(\sigma-1)}{1+\beta(\sigma-1)-\alpha\mu}\frac{L}{N} - \frac{1+\beta(\sigma-1)}{1+\beta(\sigma-1)-\mu\alpha}L_f\right] \tag{E.10}$$

Using the dynamic equation for the shadow price (equation (E.7)) and substituting equation (E.8) a second relation between the growth rate and the number of firms can be derived (based on dynamic considerations, DYN):

$$g_{DYN}^{FB} = \xi N^{\mu\alpha}\left[\frac{1}{\beta(\rho-1)(1+\beta(\sigma-1))+\alpha\mu\beta}\left(\beta\frac{L}{N}-\frac{\theta(1+\beta(\sigma-1))}{\xi N^{\mu\alpha}}\right)\right] \tag{E.11}$$

Deriving an explicit expression for the optimal rate of growth and the optimal number of firms is only possible in special cases (for example in case $\mu=1/\alpha$). In general, non-linearities do not allow us to derive explicit solutions and hence we rely on numerical techniques to analyse the interaction between the parameters driving spillovers and the optimal rate of growth and number of firms.

For the market similar expressions to (E.10) and (E.11) can be derived, but we need to use a somewhat different solution strategy than is used in the main text. Combining the zero profit condition and the demand condition governing the allocation of labour between the two sectors with labour market closure and the growth engine (equation (7.39)), yields a relation between growth and the number of firms (based on static optimality conditions, STAT):

$$g_{STAT}^{M}=\xi N^{\mu\alpha}\left[\frac{\beta}{\varepsilon}\frac{L}{N}-L_f\right] \tag{E.12}$$

A second relation between g and N based on the dynamic equation governing knowledge accumulation (7.15) and the Ramsey rule yields:

$$g_{DYN}^{M}=\xi N^{\mu\alpha}\left[\frac{1}{\alpha+\beta(\rho-1)}\left(\frac{\varepsilon-1}{\varepsilon}\beta\frac{L}{N}-\frac{\theta}{\xi N^{\mu\alpha}}\right)\right] \tag{E.13}$$

Having established the solution for g and N (in implicit form) for both the first-best social optimum and the market equilibrium, we can determine the curves in Figure 7.4 by a numerical solution procedure. The procedure used to derive the curves in Figure 7.4 is as follows: the GG-curve is found by solving for μ such that the growth rate in the market equals the growth rate in the first-best optimum for each value of (1-α). Similarly, the NN-curve is found by solving for μ such that the number of firms in the market equals the number of firms in the first-best social optimum for each value of 1-α. In the numerical experiment underlying Figure 7.4, the following parameter values were used: β=1, ε=4.7, σ=1.27, L_f=0.5, L=100, ρ=6, θ=0.01, and ξ=0.006. Notice that distortionary pricing is excluded by setting β equal to one and that the productivity effect and the business stealing effect cancel as $\sigma=\varepsilon/(\varepsilon-1)=1.27$.

NOTES

1. Combining equations (A.4) and (7.10) results in $L_x=r/\xi$. Substitution of this expression for production labour, and $L_r(=g/\xi)$ into equation (A.5) yields equation (A.6).

PART III

Empirical applications

8. Economic development and trade in the world economy: Introducing WorldScan

1. INTRODUCTION

One of the remarkable features in the world economy is the high growth rate of emerging Asian economies in the recent past. These growth rates may be explained as a catching-up phenomenon. In this view the emerging economies are imitating at a rapid pace the technological achievements of developed countries. This explanation is not unchallenged as critics have argued that rapid growth in a number of Asian economies is the result of exorbitant high savings ratios. The relative impact of capital accumulation and other determinants of economic growth can be assessed by applying the familiar method of growth accounting. It can then be shown that for some emerging countries factor accumulation leaves little room for (total) productivity improvement (for example Young, 1994; Kim and Lau, 1994).

The question has been raised of whether the high growth rates of emerging economies form a threat to developed economies. The developments in the latter group of countries have been characterized by a deterioration of the position of unskilled labour (specifically in the US) or by structural unemployment with poor perspectives for unskilled labour (predominantly in Europe) over the past decade. It is not entirely clear what the reasons are for this malfunctioning of developed economies. Some economists blame technological change, which may be biased towards unskilled labour savings (for example Krugman and Lawrence, 1994). Others think that competition from low wage countries goes far in explaining the economic problems of Europe and America (for example Leamer, 1996a; Wood, 1994). And there are authors holding a middle position (for example Freeman, 1995).

The discussion on the impact of international trade is mostly cast in terms of the standard Heckscher-Ohlin model with two factors of production, two sectors and two countries. In this chapter we intend to discuss the issue in a broader perspective by applying a calibrated model for the world economy with four regions, eight sectors and three production factors. The model called WorldScan is developed at the CPB Netherlands Bureau for Economic Policy Analysis.[1] As a stepping stone to that analysis we first consider extensions of the Heckscher-Ohlin framework proposed by Leamer (1987, 1995) and by Leamer and Lundborg (1995). The authors study the impact of the accumulation of physical and human capital in a model with homogenous goods and a number of countries with different factor endowments. It is shown that developed countries which are falling behind with respect to the accumulation of human and physical capital face severe competition from two sides. Emerging countries are gaining in the relatively labour-intensive or moderately capital-intensive sectors, while competing developed countries may become stronger in capital-intensive sectors or sectors using relatively high levels of human capital.

Accumulation of physical capital in developing regions leads in WorldScan to similar results with respect to skill premiums as in the Leamer model. This is reassuring at first sight. However, the route towards this result differs considerably in both models. What this shows is that simple models may lead readers astray by neglecting crucial elements in order to convey a clear-cut message. In complex matters one has to rely on complex models. This is the main motivation for the present chapter, which is organized along the following lines.

The Leamer model is discussed in Section 2. Applying an ingenious graphical device, developed in Leamer (1987), Rybczynski effects of capital accumulation in several regions are derived. Taking into account that massive capital accumulation also results in a relative price change of capital intensive goods, the accompanying Stolper-Samuelson effects are derived. These effects show to what extent wages of unskilled workers in rich countries are threatened by developments elsewhere in the world. Section 3 is devoted to WorldScan. The main characteristics of the model are explained in Section 3.1. The reference path which is applied in numerical experiments is presented in Section 3.2. As a first experiment we consider what would happen to other regions if the developing region were cut off. The implications of this counterfactual analysis, which show to what extent changes in skill premiums may be the result of international trade, are discussed in Section 3.3. In the second experiment, presented in Section 3.4, emphasis lies on the impact of growth in developing regions. More precisely the experiment gives the results of an increase in savings in the Non-OECD region. Section 4 closes by summing up the main results.

2. FACTOR ACCUMULATION AND CHANGING PATTERNS OF SPECIALIZATION

In the Heckscher-Ohlin trade model it is shown how domestic production factors are allocated across sectors producing tradeable commodities. The theory does not explain growth, but may be applied in a dynamic context by introducing exogenous factor accumulation. There is no specific role for technological change, because it is assumed that technological knowledge is the same in every region of the world. What distinguishes regions from each other is relative factor availability and the production mix which results from the optimal allocation of these factors of production.

The theory can be applied for any number of factors, commodities and regions, but should be kept manageable by restricting the number of factors. As shown by Leamer (1987), a stylized model with three factors and a moderate number of sectors and regions can be conveniently analysed by applying an ingenious graphical device. The three factors introduced here are labour, human capital and physical capital. The two types of capital can be accumulated over time. In accordance with the view of Kim and Lau (1994) it will be assumed that the Asian region catches up by strongly investing in physical capital. This leads to change in the production pattern of the economy towards more capital-intensive goods. Europe, America and Japan ought to give up market share in these sectors and may expand production in human capital intensive sectors, or in sectors with a very strong capital intensity. However, for this to happen these regions have to accumulate sufficient human and physical capital. As a consequence, Europe, America and Japan have to compete in the more advanced sectors in order to escape competition from Asia and other emerging economies in the world. There may be winners and losers in this respect. A region or country in the developed world may be unable to deal with increased competition from all sides and may be forced to specialize in moderately capital-intensive goods. Leamer and Lundborg (1995) refer to this possibility in what they term the Heckscher-Ohlin Crowding Hypothesis, which is applied in their paper to the recent malaise of the Swedish economy. In Leamer (1996a) the same model is applied to analyse the position of unskilled workers in America.

To illustrate the problems connected with the rise of emerging economies we follow Leamer and Lundborg by distinguishing besides the three factors of production (labour, human capital and physical capital) five sectors of production which differ with respect to factor intensity and correspond to broadly defined aggregates of two-digit industries. To account for global developments we distinguish competition among large regions in the developed world (Europe, America and Japan) on the one hand and Asia as representing emerging economies on the other hand.

The initial position of the regions with respect to relative factor availability can be shown in an endowment triangle following Leamer (1987, p. 963): 'In the three-dimension factor space, straight lines emanating from the origin contain all endowment vectors with the same ratio of factors. These lines in three dimensions can be represented by points in two dimensions by intersecting the positive orthant with a plane to form the endowment triangle. The three coordinate axes in the three-dimensional factor space are represented by the corners of an endowment triangle, and the endowment vectors of each of the countries are represented by points.' In

Figure 8.1 such an endowment triangle is presented indicating roughly relative factors availabilities around 1960. It should be observed that a point close to a vertex implies that the corresponding factor is relatively abundant. America has the most advanced economy in terms of human and physical capital. Japan and to a lesser extent Europe fall behind. Asia has a large amount of labour in comparison to human and physical capital. The example is so constructed that a line from the labour vertex through the endowment point of Europe also hits the endowment point of America.

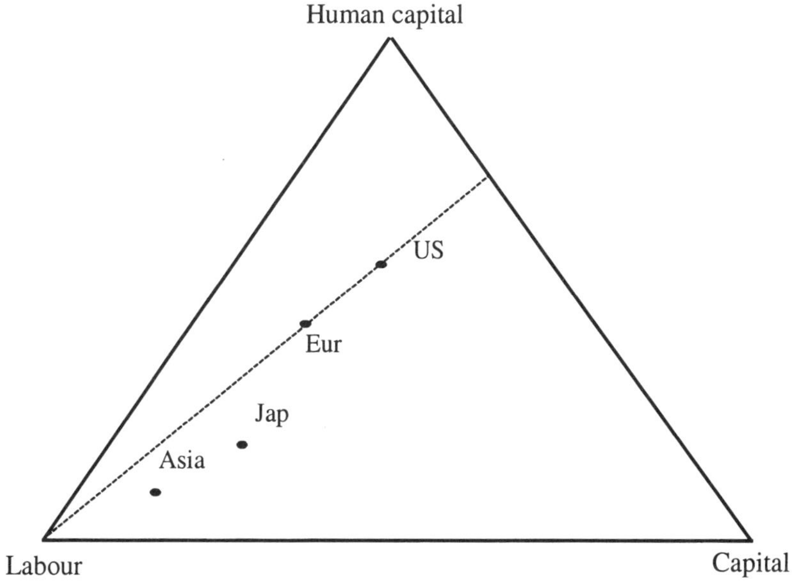

Figure 8.1 Endowment triangle

This implies that human capital and physical capital are available in the same ratio in both regions, but Europe has relatively more unskilled labour in the initial situation.

The sectors of production comprise a labour-intensive product LAB (furniture, apparel and the like), a moderately physical capital-intensive product CAP (including among others textiles, iron and steel, leather), a strongly physical capital-intensive product called machinery, MACH, and CHEM (chemical, aircraft), which is the only manufacturing sector applying human capital. This rather strong assumption is made for analytical convenience. In addition to manufacturing there is a handicraft (HAND) sector using only unskilled labour. Assuming fixed input coefficients the sectors can be represented by points in the endowment triangle as illustrated in Figure 8.2. The points indicating production activities can be connected by lines leading to triangles of diversification within the endowment triangle. In Figure 8.2 both the endowment points of regions and the activities are combined in one graph. From the linear programming literature it can be shown that countries with endowment points in a given triangle of diversification produce the three products represented by the vertices of this triangle and also have the same factor prices.

In the initial situation depicted in Figure 8.2, America is the only region producing machinery. Chemicals and the commodities represented by LAB and CAP are produced in Europe and Japan. It may be assumed that Europe imports machinery from America and exports chemicals. Japan imports chemicals and machinery in exchange for moderately capital-intensive goods (CAP) and labour-intensive goods (LAB). Asia is the only region producing labour-intensive handicrafts. Real wages are lowest in the Asian cone and increase if one moves from the left to the right. The arrows emanating from the initial endowment points show the accumulation of human and physical capital from roughly 1960 to the present.

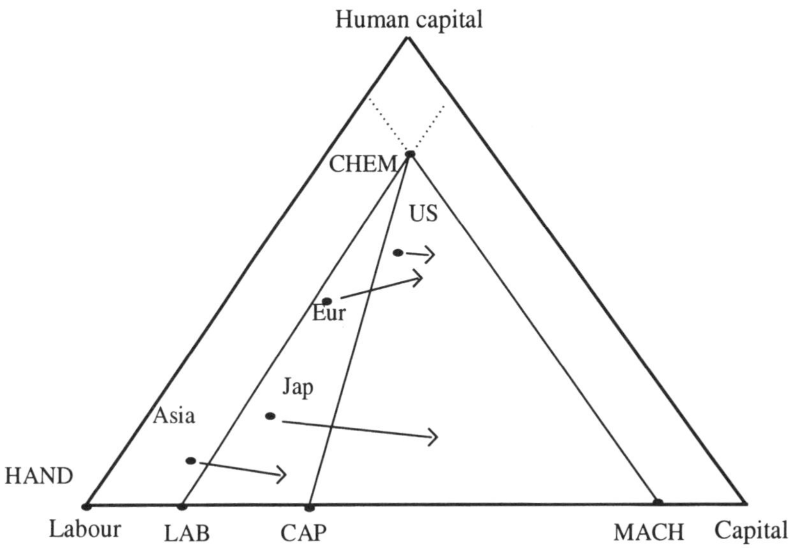

Figure 8.2 Changing Endowments

As all countries move further to the right than America does, there is convergence in the world economy. However, not all arrows are of the same length or pointing in the same directions. To unravel the implications of catching up it is necessary to determine the Rybczynski effects and the Stolper-Samuelson effects of different moves.

Table 8.1 Rybczynski effects of capital accumulation

Cone / Sector	HAND	LAB	CAP	MACH	CHEM
Asian	–	+	0	0	0
Japanese	0	–	+	0	0
American	0	0	–	+	0

Rather than going though all changes indicated in Figure 8.2 we will analyse a realistic scenario with Asia moving from the Asian cone into the Japanese cone by accumulating physical capital and Japan leaving its original cone and penetrating into the American cone also by accumulating a large amount of capital. In discussing such a scenario emphasis will be put on the consequences for the American economy. The qualitative Rybczynski effects of capital accumulation are given in Table 8.1. The effects are straightforward. It should be noted that output of chemicals does not change, because the amount of human capital is constant by assumption.

If the output of a commodity rises its price falls and the implications for factor prices follow from the Stolper-Samuelson derivatives, which are dual to the relevant Rybczynski effects. The consequences of price changes for real factor earnings in the American cone are shown in Table 8.2. Handicrafts are left out because it is assumed that they are not consumed in America.

Table 8.2 Stolper-Samuelson effects for America

Output price reduction	Factor reward	Wages unskilled	Wages skilled	Capital
LAB		0	0	0
CAP		–	+	+
MACH		+	+	–
CHEM		0	–	0

The signs relate to changes in factor rewards caused by price reductions. A fall in the price of LAB does not change factor incomes in the American cone, as the labour-intensive commodity is not produced in America. However, all factors gain in real terms due to an improvement in the terms of trade. A decline in the price of CAP, the moderately capital-intensive good (textiles, and so on), induces a fall in the wage of unskilled labour, but a rise in the earnings of skilled labour and capital. In the terminology of Ethier (1984) CAP is a 'friend' of unskilled labour, which appears from the fact that labour and CAP are on the same side of the extended line connecting machinery and chemicals in Figure 8.2. Human capital and physical capital are on the other side of the line and are therefore 'enemies' of CAP. A price reduction in this good induces an increase in the factor rewards of the 'enemies', as shown in Table 8.2. In a similar way it can be argued that capital is a 'friend' of MACH, but that human capital and labour are 'enemies'. A fall in the price of machinery leads to a decline in the remuneration of capital, but the wage rate of skilled as well as of unskilled workers rises. Finally, a lower price of chemicals has only consequences for the wage of skilled workers, which will be reduced. Capital and labour are on the extended line connecting CAP and MACH and are neither 'friends' nor 'enemies'. The results depend of course upon the assumptions made with respect to production technologies: 'Conclusions regarding the effect of product-price changes on factor returns depend substantially and subtly on the input intensities. For example, if the machinery point is swung to the left, to a capital-to-labour ratio less than that of chemicals, then human capital and textiles become friends in the capital abundant countries' (Leamer, 1995, p. 24).

Turning to the accumulation scenario mentioned above we see that the Japanese growth induces a fall in the price of machinery. This puts the rate of return on capital under pressure in America. The wages of skilled and unskilled workers are favourably affected. The Japanese expansion reduces the supply of CAP, but this is likely to be more than offset by the assumed capital accumulation in Asia. The price of CAP may decline resulting in lower wages for the unskilled workers. On balance the skill premium rises and the rate of return on capital declines, which may depress savings and capital accumulation in America. As these developments took place the Heckscher-Ohlin Crowding Hypothesis may have been relevant for the US in the recent past. The change in the distribution of wages at the expense of the low skilled is widely documented (cf. Bound and Johnson, 1992). The low savings rate is another well-known feature of the American economy. Moreover, there has been a change from a position with positive net exports per worker of machinery in 1974 to a position of importing machinery in 1988 (for example Leamer and Lundborg, 1995).

The Leamer model can be extended by discussing separately capital mobility[2] and the impact of non-traded goods. Nevertheless the model gives a highly stylized description of the world economy, mainly because goods are homogenous so that the resulting patterns of specialization are rather rigid. For a sketch of possible developments it may be preferable to rely on a calibrated model providing a more comprehensive picture of the world economy. In the next section we apply the WorldScan model of CPB Netherlands Bureau for Economic Policy Analysis to study the consequences of trade with and factor accumulation in emerging economies.

3. FUTURE DEVELOPMENTS: SOME SIMULATION EXPERIMENTS

In theory, models can be kept simple to focus on a particular issue. Different models can then be constructed for different purposes. In a calibrated model for the world economy all relevant mechanisms have to be taken into account. As a consequence, such models will be large and complicated. Presentation of the full model is beyond the scope of this chapter. Therefore, in Section 3.1 we present an overview of the model explaining the main features. The model can be used to generate future developments in terms of the endogenous variables. The results of this exercise should not be interpreted as predictions. In Section 3.2 we discuss a possible development which serves as a reference path for two experiments. A counterfactual analysis de-linking the non-OECD region from the rest of the world is presented in Section 3.3. This experiment provides useful insights with respect to the impact on developed economies of trade with emerging economies. In Section 3.4 the role of capital accumulation in the non-OECD region is studied by assuming an increase in savings in comparison with the reference path.

3.1 WorldScan

WorldScan is a multi-region multi-sector model generating numerical results, which may be useful in scanning the future. We limit the exposition to a summary statement of the main mechanisms in the model (some details are presented in Box 8.1). The model deviates from standard applied general equilibrium models by allowing markets to clear gradually by means of a combination of stock adjustment and price adjustments. In addition, there are other frictions built into the process of factor allocation. Capital as a factor of production is not fully malleable. In fact production functions are of the putty semi-putty variety. In each period newly installed capital (new vintage) is distinguished from existing capital (old vintage). Substitution possibilities on the new vintage follow from a Cobb-Douglas technology, whereas substitution possibilities on old vintages are more limited. WorldScan contains two separate labour markets, one for low-skilled and the other for high-skilled workers. In both markets people may be unemployed, because firms use the wage rate as a screening device with regard to the number of applicants for jobs. Unemployment benefits do have an impact by raising the reservation wage and therefore the number of applications by workers.

Economic growth is caused by factor accumulation and technological change. Labour supply of skilled as well as of unskilled workers is determined exogenously. Capital accumulates by investing in new capacity. Because financial capital is internationally mobile, to a certain extent, domestic saving rates may deviate from domestic investment rates.

BOX 8.1 THE WORLDSCAN MODEL

WorldScan is a multi-sector, multi-region, dynamic applied general equilibrium model for the world economy, designed to support economic scenario studies on a time frame of 25 to 35 years. It consists of a neoclassical core model for long-term development, based on a Solow growth model and a Heckscher-Ohlin trade model.

Regions	*Sectors*	*Production factors*
Western Europe	Agriculture and food (A)	Low-skilled labour (L)
Japan	Raw Materials and energy (RM)	High-skilled labour (H)
North America	Intermediate goods (I)	Capital (C)
Non-OECD	Consumption goods (Co)	
	Capital goods (Ca)	
	Non-tradables (sheltered sector) (S)	*Intermediate inputs*
	International transport services (IS)	Raw materials and energy (R&E)
		Intermediate goods (I)

Extensions to the core model
Extensions are made to add realism to scenarios and in so doing bridge the gap between academic and policy discussions. These include

- a Keynesian closure with endogenous investment introduced to illustrate medium-term dynamics of shocks;
- catching up of productivity levels to a specified technological frontier, depending upon a number of endogenous factors;
- an Armington trade specification, allowing market power to determine medium-term trade patterns, while allowing Heckscher-Ohlin mechanisms in the long run;
- imperfect financial capital mobility with a portfolio model, with a specification allowing penetration in financial markets without continuously increasing interest rates;

Extensions to the core model (cont.)

- a consumption demand system depending upon per capita income, and developing towards a universal pattern, and
- a Lewis-type informal sector in developing regions, which provides a surplus of labour supply, enabling economic growth of low-wage countries during a longer period.

The model is calibrated on a 1990 database containing consistent bilateral trade flows and input output information, with no rest of world region. Elasticities are guesstimates - no attempt has been made to estimate them using econometric methods.

Source: Geurts, Gielen, Nahuis, Tang and Timmer (1995).

Technological change is modelled by distinguishing a notional technology and the actual or applied technology in each sector and each region. The gap between the two constitutes a learning potential. Firms are chasing the notional technology but are never able to close the gap completely. In WorldScan the ability to learn is related to the capital-labour ratio and to the employment of high-skilled labour relative to the employment of low-skilled labour, thus accounting for views expressed in the recent, so called endogenous, growth theory. The notional technology of a sector is related to the actual technology in this sector, but also to the actual technology of similar sectors in other regions. In this manner international knowledge spillover-effects are taken into account. Regions that have backward technologies for whatever reason may catch up by exploiting a relatively large learning potential.

Sectors in WorldScan are not based on a straightforward aggregation of statistical categories, but are conceived as prototypes of main economic activities. Agricultural products and raw materials are traded in global markets under perfect competition. The international services sector provides for international transport services, which are traded in the world market. The products of the sheltered sector are supplied in regional markets only. The three remaining manufacturing industries supply products on regional markets, that are open to interregional trade. Preferences are modelled according to an adjusted version of the Armington system. Whereas in the original Armington system market shares are determined by relative price levels, in WorldScan *changes* in market shares depend upon relative price *levels*. As a consequence, substitution possibilities in the long run are larger than those in the short run.

The capital goods sector can be thought of as a high-tech sector, using a relatively large amount of skilled labour and also more physical capital per unit of output than the other manufacturing sectors: consumer goods and intermediate goods. The consumer goods sector is a more traditional sector applying a relatively large amount of unskilled labour. The intermediate goods sector, of which the goods are used in the production of all other goods, holds a middle position with respect to capital-intensity and skill-intensity. The production elasticities which are indicative for the relative factor intensities of the sectors are presented in Table 8.3. In assessing the relative factor intensities of sectors account should be taken of the indirect effects of the input-output relations.

Except for agricultural and consumer goods the elasticities with respect to the use of intermediate products are equal and low. Raw materials are used in the intermediate goods sector only, where they take a substantial share of gross earnings.

Table 8.3 Production elasticities in WorldScan

Production factor[a] Sector	H	L	C	I	R&E
Capital goods	0.45	0.15	0.30	0.10	0.00
Sheltered	0.40	0.20	0.30	0.10	0.00
Transport	0.40	0.20	0.30	0.10	0.00
Raw materials	0.35	0.35	0.20	0.10	0.00
Agriculture	0.15	0.45	0.15	0.25	0.00
Intermediate goods	0.15	0.15	0.20	0.00	0.50
Consumer Goods	0.15	0.50	0.20	0.15	0.00
Informal Sector	0.00	0.80	0.10	0.10	0.00

Note: [a] The production factors' abbreviations are defined in Box 8.1.

The numerical accuracy of the model requires some discussion. The model is calibrated such that it exactly replicates a 1990 database. The database is constructed using several sources, see Geurts et al. (1995). No database with sufficient sectoral detail covering the globe with a reliable time series dimension is available, to our knowledge. Hence, the disadvantage of having only one data-point seems unavoidable. Given this limitation, the numerical results produced with the model should not be interpreted as predictions based on an econometrically estimated model. Rather, the results are indicative answers to 'what if' type of questions with a *reasonable* order of magnitude.

First, the model will be applied to gauge the effect of trade with the non-OECD region on OECD labour markets. Next, the model will be applied in analysing the impact of an increase in savings in the non-OECD area. It remains to be seen to what extent the Heckscher-Ohlin pattern is preserved in a more realistic setting. The shocks referred to above are applied to a reference path of the model for the period 1995–2025, which is briefly discussed in the next section.

3.2 The reference path

Although we do not intend to predict future developments one way or the other it may nevertheless be useful to discuss some main characteristics of the reference path.

In Table 8.4(a), (b) and (c) we report macroeconomic growth and its determinants for four main regions in the world economy. The developed regions show a more or less similar growth pattern, although labour supply stagnates in Japan. The difference in Europe and America is made up by faster accumulation of physical capital. Employment in the non-OECD area grows faster than labour supply, which is made possible by shifting workers from the informal sector to the formal sector. The savings rate as a percentage of national income amounts to 24 per cent on average, which is substantially less than Japanese saving of about 35 per cent of national income.

Table 8.4(a) Reference path 1995–2025

	Western Europe	North America	Japan	Non-OECD
	determinants of income growth (% p.y.)			
Labour supply	0.5	0.7	0.0	1.3
Inflow inf. sector				0.4
Factor productivity	0.6	0.9	0.9	1.3
Factor accumulation	0.7	0.7	1.3	1.1
National product	1.9	2.3	2.1	4.0

Table 8.4(b) Reference path 1995–2025

	Western Europe		North America		Japan		Non-OECD	
	regional share in world economy (%)							
	2000	–25	2000	–25	2000	–25	2000	–25
Agriculture	27.4	19.9	18.4	14.4	8.2	5.5	46.0	60.2
Raw materials	22.7	23.8	33.6	37.6	5.8	4.5	38.0	34.1
Int. services	27.4	30.5	20.0	22.5	18.0	16.0	34.6	30.9
Sheltered	36.5	30.6	31.6	32.1	17.5	13.6	14.4	23.7
Interm. products	33.1	25.6	28.0	25.6	14.5	9.9	24.4	38.9
Cons. goods	14.1	4.7	8.3	4.7	14.4	6.8	63.3	83.3
Capital goods	35.0	26.7	27.7	31.5	25.2	20.0	12.1	21.8

Knowledge spillovers, triggered by capital accumulation and schooling, lead to catching up in terms of total factor productivity as appears from Table 8.4(a).

As the non-OECD is growing faster than the rest of the world its share in world production goes up in most sectors. Exceptions to the rule are raw materials and international services. However, there are no big changes in the shares of local income. In the non-OECD region there is a slight increase in the shares of tradables originating in industry at the expense of raw materials mainly. Note, for future reference, that the world division of labour in manufacturing sectors is characterized by a specialization of the non-OECD on consumption goods and of the OECD on the capital goods sector. This is revealed both by shares in the world economy and by shares in local income.

Table 8.4(c) Reference path 1995–2025

	Western Europe		North America		Japan		Non-OECD	
	sectoral share in local income (%)							
	2000	–25	2000	–25	2000	–25	2000	–25
Agriculture	8.5	8.5	6.5	5.8	5.3	5.2	19.6	20.3
Raw materials	4.6	5.6	7.7	8.3	2.3	2.3	9.7	6.4
Int. services	0.7	0.6	0.6	0.4	1.0	0.7	1.1	0.5
Sheltered	65.7	66.4	66.3	65.9	64.7	65.1	38.2	40.3
Interm. products	8.2	7.5	7.4	7.1	7.7	6.4	7.7	8.6
Cons. goods	1.9	1.0	1.3	0.9	4.4	3.1	12.9	13.2
Capital goods	10.4	10.4	10.3	11.5	14.7	17.1	5.5	6.6

All in all economic growth is rather balanced with developing regions converging over a broad spectrum of the economy. Nevertheless, at the end of the simulation period the non-OECD countries are still markedly different from the developed economies as appears from a relatively large agricultural sector, a relatively small sheltered sector and a relatively large consumption goods sector.

3.3 De-linking the non-OECD: A counterfactual analysis

By means of a counterfactual analysis we assess the points of view expressed in the introduction with respect to the deterioration of the position of unskilled labour. By determining the fraction of increased wage inequality that *is* explained by trade with the non-OECD, we evaluate the validity of arguing that it is predominantly a domestic technology related phenomenon, predominantly a trade related phenomenon or some mixture of the two. To assess the position of those advancing the trade explanation in the wage inequality debate, the relevant question can be formulated as: 'Is it possible to imagine an alternative history in which the OECD countries have acquired the technology and resources of the mid-1990s, but in which trade with newly industrialising countries remains negligible (either because these countries did not develop, or because protectionist barriers have blocked off the potential trade)?' (Krugman 1995b, p. 22). The impact on the OECD labour markets of an alternative history without trade between the developed and developing countries is analysed in this section. Hence the experiment discussed in this section can be conceived as accepting Krugman's challenge: 'At the very least, I would challenge those economists who maintain that trade has in fact had large impacts on wages to produce an illustrative CGE model, with plausible parameters, that is consistent with that view. I claim that it cannot be done' (Krugman 1995a, p. 22).[3]

The layout of the de-linking of the non-OECD scenario is as follows. The non-OECD region raises drastic trade barriers with the aim of becoming isolated from the rest of the world economy. On all (potential) trade flows import and export taxes are levied in order to squeeze the actual trade flows as quickly and as far as possible from 1991 onwards. Technological change is exogenous at the reference path level in order to focus on the Heckscher-Ohlin trade effect. Tax revenues are returned as a lump sum payment to the representative consumer.

The results are reported in Table 8.5.[4] Our focus is on the labour markets in the OECD regions. In Western Europe, North America and Japan the wages of low-skilled workers rise whereas those of high-skilled workers fall. Hence, *de-linking* the non-OECD brings about a decrease in the wage gap in the OECD. The significance of the magnitude of the relative wage change is discussed at the end of this section; first the results will be explained.

The simulation results reflect basically the analytical results of the Stolper-Samuelson theorem. The price of the low-skilled intensive imports rises relative to the high-skilled intensive exports. The logic of the Stolper-Samuelson theorem is that to maintain zero profit the wage of the low skilled should increase and the wage of the high skilled should decrease.[5] This is what we observe in the OECD regions.

Trade with the non-OECD is eliminated by trade barriers erected by the non-OECD region. The trade pattern is not shown, but the fact that trade is negligible can be inferred from the change in the share in local income of the international (transport) services in the non-OECD: a decrease with 0.7 percentage-points from a reference path value of 0.8 per cent in 2005. Without international trade there is no demand for international transport services. Recall that in the reference path (see Table 8.4) the non-OECD is the dominant producer of consumer goods. The structure of trade on the reference path mirrors this specialization pattern. The non-OECD exports predominantly consumer goods (and raw materials) which balance the import of goods from the high-skilled intensive capital goods sector. In this experiment the tariffs block off this trade flow. Therefore the cheap (non-OECD) consumption goods are no longer available for the OECD nor is there a demand for capital goods from the non-OECD. This leads in the OECD to an increase in the consumption goods price index and to a decrease in the capital goods price. This change in relative prices is accompanied by a sectoral reallocation.

Table 8.5 De-linking the non-OECD (Cumulative effects in %)

	Western Europe 2005	North America 2005	Japan 2005	non-OECD 2005
		Real factor prices		
Wage low skilled	2.3	2.1	3.4	-3.1
Wage high skilled	-2.3	-2.9	-4.3	12.5
User cost capital	-2.5	-3.6	-0.4	0.5
		Skill premium[a]		
Skill premium base	1.55	1.61	1.58	2.99
Skill premium delink	1.48	1.53	1.46	3.47
Difference	-0.07	-0.08	-0.12	0.48
		Shares in local income		
Agriculture	0.6	0.7	0.9	-0.2
Raw materials	0.3	1.0	0.6	-1.6
Int. services	-0.1	-0.1	-0.1	-0.7
Sheltered	-0.4	-0.6	-0.8	0.4
Intermed. products	-0.2	0.1	-0.1	0.5
Cons. goods	1.7	1.1	1.8	-3.7
Capital goods	-2.1	-2.1	-2.4	5.1
		Revealed comparative advantage		
Intermed. products	3.4	10.8	3.1	..[b]
Cons. goods	24.1	18.1	15.8	..
Capital goods	-8.8	-9.6	-6.1	..

Notes: [a] Wage high skilled / wage low skilled in absolute value.
[b] The computation of RCA is not informative if trade volumes tend to zero.

It is apparent from Table 8.5 that the consumption goods share in local income goes up at the expense of the share of capital goods in all OECD regions. The share of raw materials also rises because of the elimination of the non-OECD exports. International services is the only sector that decreases in local income share in all four regions, this reflects the fact that total trade flows have diminished considerably. In all other sectors (with the exception of the minor changes in the intermediate goods sector) the pattern in the non-OECD is opposite to that of the OECD regions. Hence the changes in labour demand in the non-OECD also show the opposite pattern of that in the OECD.

What remains to be explained is the decrease in the user cost of capital. This reflects the fact that capital consists of goods from the sheltered sector (buildings) and from the capital goods sector, both are skilled intensive and hence are produced cheaper due to the decline in high-skilled labour wage costs.

How do the labour market results relate to the actual changes observed in the OECD economy? Davis (1992) provides numbers for the ratio of payment of workers having finished college to those who finished only high school. For the US this ratio increased by 0.15 points from 1.37 in 1979 to 1.52 in 1987. For the UK the ratio increased by 0.12 points from 1.53 to 1.65 in eight years (1980–88).[6] Table 8.5 depicts the absolute values for the skill-premia (the wage of high-skilled workers relative to low-skilled workers) in the reference path and in the *de-linking* simulation. For Western Europe the skill premium decreases by 0.07 points from 1.55 in the reference path to 1.48 in the *de-linking* simulation. Comparing this number, which reflects the full impact of non-OECD trade, with the actual number for the UK of 0.12 one might say that half of the increase would be undone had trade with low wage countries not been possible. For North America the actual increase in 8 years of 0.15 points compares to a full simulated impact of *de-linking* of 0.08 points. Hence, the counterfactual experiment learns that the non-OECD might be 'blamed' for somewhat less than half of the increase in the skill premium in the OECD. This is not to say that the impact of trade on OECD labour markets has been large, but it has been too large to argue that the changing skill premium is predominantly a domestic technology related phenomenon.

3.4 Increasing the saving propensity in the non-OECD region

The cumulative results of a rise in the saving ratio in the non-OECD region are given in Table 8.6. The shock amounts to a permanent increase by 10 percentage-points starting in the year 1995. The results for the non-OECD region will be discussed first. The user cost of capital in this area decreases substantially, which induces a rise in the production of capital intensive goods (sheltered, capital goods and international services). The share of these goods in local income goes up and the revealed-comparative-advantage measure in the table points towards an improvement in the competitive position of the capital-goods sector in the world market. This reflects the Rybczynski theorem with an increased capital endowment. The increased capital endowment does not fall from heaven, consumers save more (and spend less) and the capital stock has to be produced. In the short run an increase in savings induces underspending and unemployment, as the model exhibits Keynesian features. As a consequence unemployment increases and factor remunerations decline. In the long run the natural rate of unemployment is restored and changes in labour income are governed by the required reallocation of labour from the relatively less skill-intensive sectors. As appears from Table 8.6 the decrease in consumption lowers demand for consumption goods and for agricultural products. The production of capital goods and the supply of goods from the sheltered sector, which are used in construction (and are therefore part of total investment) rise. The decline in the user cost of capital makes it possible for wages to rise over the full range. However, wages of skilled workers rise more because there is a reallocation towards the skill-intensive sectors (capital goods and sheltered).

The figures for revealed comparative advantage indicate that the non-OECD area will specialize more in capital goods. However, this does not imply that this region will export more or import less of these goods as is assumed in the Leamer model discussed in Section 2. The extra demand for domestic capital goods raises the price of these goods, so that it becomes even more attractive to import capital goods from the OECD regions. An increase in investment in the non-OECD region increases the world demand for capital goods.

In the short run the OECD regions face a substantial rise in the price of capital goods instead of a decline as would be the case in the Leamer analysis if accumulating countries penetrate in the American cone and start to produce machinery. In the long run the relative price of capital goods rises in comparison with agricultural goods and products from the sheltered sector. The other relative prices do not change much. According to the Stolper-Samuelson theorem an increase in the price of the relatively skill-intensive capital good induces a rise in the remuneration of skilled labour in the OECD regions. This result is confirmed by the outcomes presented in Table 8.6. However, this is not the whole story. The user cost of capital declines because there is (imperfect) capital mobility in the model. The fall in the non-OECD rate of interest spills over to the OECD regions. The resulting decline in the remuneration of capital improves the position of labour. Skilled labour benefits most, because the relatively capital-intensive goods are also the most skill-intensive goods. If the relative price of capital does not change the fall in capital cost must be compensated for by a wage increase of skilled workers. With the exception of Japan the wage rate of unskilled workers also rises in the long run. In Japan the user cost of capital decline less than in the other OECD regions.

The simulation experiment of an increase in capital accumulation in the non-OECD area produces results which deviate in a certain way from the standard Heckscher-Ohlin theory. Account must be taken of supply as well as of demand effects. The need for more capital goods in the non-OECD will be partially met by importing these goods from the rest of the world. This leads in the medium run to a price increase of capital goods world-wide in all regions. Moreover, capital mobility induces spillovers with respect to the income distribution.

Table 8.6 Increase in non-OECD savings (Cumulative effects in %)

	Western Europe 2000–25		North America 2000–25		Japan 2000–25		Non-OECD 2000–25	
				Real factor prices				
Wage Low	-1.1	0.1	-1.6	0.2	-1.0	-0.3	-2.9	6.6
Wage High	0.8	1.1	0.3	1.6	2.4	1.3	-3.4	15.0
Cost Cap.	-1.1	-1.4	-1.6	-1.9	-1.3	-0.7	-13.5	-27.9
				Shares in local income				
A	-0.9	-0.4	-0.7	-0.4	-0.9	-0.4	-1.5	-1.7
RM	-0.2	0.1	-0.5	-0.1	-0.3	0.0	-0.1	-0.3
IS	0.3	0.1	0.2	0.0	0.4	0.0	0.5	0.1
S	0.2	-0.3	0.0	-0.5	0.5	-0.1	2.2	2.0
I	-0.3	-0.1	-0.1	-0.1	-0.4	-0.1	-0.6	-0.3
Co	-0.1	-0.1	-0.1	-0.1	-0.2	-0.2	-1.8	-1.1
Ca	1.1	0.7	1.1	1.0	0.9	0.9	1.8	1.6
				Revealed comparative advantage				
I	-1.4	1.1	0.4	-1.2	-0.5	1.2	0.4	0.1
Co	2.2	-0.8	0.9	-1.4	6.3	-0.9	1.9	-4.3
Ca	-4.4	-4.5	-6.5	-3.8	-7.0	-7.3	11.0	9.0

Note: Definitions for the abbreviations are provided in Box 8.1.

4. CONCLUSIONS

Trade with emerging economies is often blamed for the relative or absolute deterioration of the income position of low skilled workers in developed economies. Applying the WorldScan model of the CPB Netherlands Bureau for Economic Policy Analysis, it is shown that about half of the increase in skill premium in the US and in the UK can be attributed to trade with the non-OECD region. This result, which is an upper bound, should be interpreted with great caution because the model is not estimated by applying econometric methods. As a consequence of limited data availability the model is merely calibrated as is the case with most applied general equilibrium models. Nevertheless it is reassuring that the result is in line with empirical work showing that at least a substantial part of the increase in the skill premium can be attributed to trade with emerging economies.

From a dynamic perspective changing patterns of trade are determined by unequal growth across regions. A number of emerging economies have shown high rates of output growth over the past decades. Recent empirical investigations have shown that growth in emerging economies is driven to a large extend by capital accumulation. This result makes it possible to apply the time-honoured Heckscher-Ohlin model in analysing the issue of changes in the skill premium.

If capital increases relatively fast in emerging economies, one would expect that these economies increase their output and export in moderately capital intensive goods. As a result the price of these goods in developing as well as in developed countries would fall. Moderately capital intensive goods are at the same time relatively skill extensive, so that a fall in the price leads to a reduction in the wage rate of unskilled workers according to the Stolper-Samuelson theorem. This is indeed the result found in Leamer (1995, 1996a) in a model with five goods, three factors of production and four regions.

In contrast with Leamer, in WorldScan goods from different regions are heterogenous. Consumers have a taste for variety, so that all goods are produced in all regions. This is a logical assumption in a calibrated model with a limited number of sectors. Capital accumulation in the non-OECD region is conceived as an increase in the savings rate. Demand for capital goods rises at home as well as abroad because the import of capital goods in the non-OECD area increases. In the long run the increase in demand can be met at more or less stable relative prices in the OECD regions. User costs of capital fall in the non-OECD area, but also in other regions as financial capital is imperfectly mobile. The decline in capital cost is beneficial to labour. Skilled labour benefits most, because the production of capital goods is relatively capital intensive as well as relatively skill-intensive.

It is interesting to observe that the Leamer analysis and WorldScan lead to a similar conclusions with respect to the skill premium in the case of capital accumulation in emerging economies. However, the road leading toward this conclusion is very different. This may justify the use of more complex models despite the fact that no closed-form solution for such models can be found.

NOTES

1. For a description of the model see Geurts et al. (1995).
2. Capital mobility leads to a change in endowment points. As capital moves from America to Asia competition increases. This may put additional pressure on the wages of unskilled labour in America eventually.
3. See Nahuis (1999) for a more extensive motivation on why this approach and the use of WorldScan is particularly useful.
4. The full effect of this experiment is attained in 2005. In WorldScan slowly vanishing brand loyalty is introduced such that differences in prices of regionally differentiated varieties of a good are translated gradually in changing market shares (see also Section 3.1). In the end markets can be lost completely, however it takes till 2005 to squeeze non-OECD trade almost completely.
5. A different way to understand the tendency for the relative wage to move in a direction beneficial for those with less skills is to see the world as a (single) closed economy. After tax walls have been raised, the relative endowment of the closed world economy consisting of Europe, Japan and North America, is such that unskilled workers are much more scarce than with the non-OECD included.
6. A more extended overview of labour market trends in the 1980s is presented in Kosters (1994).

9. Openness, growth and R&D spillovers: An R&D-amended version of WorldScan

1. INTRODUCTION

Are countries with lower barriers to trade experiencing more economic progress? Trade economists typically answer this question affirmative despite the fact that neo-classical trade theory predicts that lower barriers to trade will lead to higher *levels* of welfare only (as long as a country is small). The Solow growth theory predicts no link between trade barriers and growth whatsoever. Only in the transition phase openness might have an effect on growth. Models of endogenous growth provide the 'missing link' between openness and growth.[1] Openness has growth effects via knowledge spillovers related to openness that affect the productivity of research or production, or reduce duplicationary research effort. Openness can also allow to benefit from specialization (or scale) opportunities in research or generate a market-size effect.[2]

Coe and Helpman (1995) have quantified directly the relation between technological change, openness and research expenditures within the OECD. They have shown that R&D is not only beneficial for the performing countries, but also for their trade partners.[3]

This chapter integrates the empirical results and the theory in an Applied General Equilibrium (further AGE) model. We examine the importance of R&D and R&D spillovers in quantifying the effects of trade liberalization.[4] We do so in two steps. First, we estimate an equation related to Coe and

Helpman (1995). The results are subsequently implemented in WorldScan, an AGE model for the world economy with considerable sectoral detail. Finally, the consequences of trade liberalization are considered.

Closely related is the work by Bayoumi, Coe and Helpman (1999, further BCH) who implement the estimated equation of Coe and Helpman (1995) in the dynamic multi-country model of the IMF (MULTIMOD). They show that a trade expansion by developing countries of 5 per cent of their GDP raises their output by 6.5 percentage-points in 2075.[5]

The analysis in this chapter adds to BCH's paper in several respects. First, given that we have sectoral detail in the model, we can distinguish intra-regional spillovers alongside inter-regional spillovers. Second, we collect and incorporate R&D data for non-OECD regions whereas BCH's assumptions imply that these regions do not perform any R&D till 2075. Third, to highlight the role for trade as a vehicle for R&D spillovers we perform a different exercise to BCH. We argue that a relevant policy shock is to reduce existing trade barriers over time, whereas BCH increase imports and exports of manufactures by 5 percentage-points. Finally, we distinguish high- and low-skilled workers. This allows us to examine the hypothesis, propagated by Wood (1994), that trade liberalization causes 'defensive' innovations that are harmful for low-skilled workers. These differences in the approaches immediately allow us to pin down the points this chapter makes.

First, we show that trade-related R&D spillovers not necessarily magnify the effects of trade liberalization. This is related to our different perspective on the relevant trade-liberalization experiment. We introduce trade liberalization starting from the trade barriers in the data. This implies that the relative prices of the regional varieties are affected by liberalization. This may redirect trade flows and thereby affect the 'imported' knowledge flows. This results (for some regions) in very low benefits from international spillovers as they import less knowledge-intensive products. The BCH experiment veiled this as they increased the import intensity in a neutral way.

The second point this chapter makes is that it is crucial to distinguish intra-national spillovers alongside international spillovers as it brings to the fore the trade-off between the two. This point is easily understood once the trivial observation, that goods that are imported are not produced domestically, is recognized. Note that a large market induces more R&D. Hence, increased 'imported' international spillovers come at a cost of domestically generated knowledge (which might be important for intra-national spillovers).[6] Hence, trade liberalization might cause regions to specialize in sectors that have low growth potential.

The third point of the chapter is that there is no evidence for a technology-related magnification effect on the relative wages of high- and low-skilled workers. The intuition for a magnification effect on relative wages in the OECD goes as follows. Trade liberalization leads to a lower relative price of unskilled-intensive goods. This leads to a sectoral reallocation towards skilled-intensive industries, which leads to higher R&D expenditures in these industries, finally leading to a sector bias in TFP. We do not find such an effect. There are three reasons for this. First, the sensitivity of TFP for 'own'-sector R&D is low. Second, some regions do not specialize in high-skill/R&D-intensive sectors. And third, the high-skill/R&D-intensive sectors generate considerable spillovers within the domestic economy towards low-skilled intensive sectors.

The rest of this chapter is organized as follows. In Section 2 we discuss our empirical model. The estimation results for this model are presented in Section 3. This section also contains a discussion of the data. Section 4 presents WorldScan, the AGE model. Section 5 presents our main results. Section 6 concludes.

2. THE EMPIRICAL MODEL

There is a substantial literature, both theoretical and empirical, that relates R&D expenditures to productivity growth. The view that technological progress benefits not only from R&D performed within the sector, but also from R&D performed 'elsewhere' is well established (see Nadiri, 1993, for an overview of the literature). More recently, inspired by endogenous growth theory, the link between productivity and R&D performed in other countries has been emphasized in empirical work.[7] This section sets out a model to re-examine these intra- and international spillovers.

2.1 Empirical approach

We start the analysis with a production function for a specific sector in a country:

$$Y = G(A, Q, L, R^{DD}, R^{ID}, R^{F}) \ , \tag{9.1}$$

where, Y, A, Q and L denote value added, technology, capital and labour respectively. The Rs denote weighted knowledge stocks (that are a function of R&D expenditures). The superscripts to the Rs have the following meaning: DD is Direct (same sector) Domestic and ID Indirect (other sectors) Domestic. The superscript F should be read as Foreign. Time differentiation and bringing the terms with L and Q to the left-hand side yields:

$$\begin{aligned} &\hat{Y} - (\alpha_L \hat{L} + (1-\alpha_L)\hat{Q}) \equiv \\ &\hat{F} = \Psi_A \hat{A} + \Psi_{R^{DD}} \hat{R}^{DD} + \Psi_{R^{ID}} \hat{R}^{ID} + \Psi_{R^F} \hat{R}^{F} \ , \end{aligned} \tag{9.2}$$

where $\Psi_i = \partial G/\partial i \bullet i/Y$ for $i \in [A, L, Q, R^{DD}, R^{ID}, R^{F}]$ and a hat over a variable denotes a growth rate. Assuming competitive markets and constant returns to scale in labour and capital[8] we can infer Ψ_L and Ψ_Q from the data as the factor-income shares (denoted as α).[9] F is the TFP level.[10]

We apply a growth-accounting procedure to capital and labour and estimate for the R&D impact, instead of estimating the complete production function or assessing the impact of R&D also by means of growth accounting. We do so for the following reasons. First, one might argue that by accounting for TFP growth the R^{DD} variable should be included on the left-hand side. However, these expenditures are included in capital and labour already (see also note 9). Second, growth accounting for Q and L

overcomes the problem that – common in the empirical literature – strongly decreasing returns are found (see, for example, Verspagen, 1997), which are due to measurement errors in the capital stock. Barro (1999, p. 122-3) is more extensive on this issue. Finally, the exogeneity of the growth rates of Q and L with respect to technological change is doubtful.[11]

We prefer to incorporate in the AGE model the following specification that is closely related to R&D-based endogenous growth models[12]

$$F_{ik} = A_{ik}\left(R_{ik}^{DD}\right)^{\gamma_{ik}^{DD}}\left(R_{ik}^{ID}\right)^{\gamma_{ik}^{ID}}\left(R_{ik}^{F}\right)^{\gamma_{ik}^{F}}, \tag{9.3}$$

C&H (1995) use a similar specification. F_{ik} is the TFP level for industry i in region k (industries are denoted by i ($i \in \{1,...I\}$) and regions by k ($k \in \{1,...K\}$). The superscripts to the Rs have the following meaning: DD is direct (same sector) domestic and ID indirect (other sectors) domestic. The superscript F should be read as foreign.

We estimate the time derivative of equation (9.3), which is in line with equation (9.2):

$$\hat{F}_{ikt} = c + \gamma_{ik}^{DD}\hat{R}_{ikt}^{DD} + \gamma_{ik}^{ID}\hat{R}_{ikt}^{ID} + \gamma_{ik}^{F}\hat{R}_{ikt}^{F} + \varepsilon_{ikt}, \tag{9.4}$$

where a constant c captures the unexplained exogenous growth trend. An error term is added.

We follow Grossman and Helpman (1991) in the logic that research productivity and thus productivity growth depends on the knowledge stock available for R&D. Therefore, knowledge stocks are weighted *sums* of other sectors' and countries' R&D-capital stocks.

$$R_{ik}^{ID} = \sum_{j \neq i}^{I} \omega_{jik}\eta_{jik}R_{jk} \qquad R_{ik}^{F} = \sum_{l \neq k}^{K}\sum_{j}^{I} n_{jilk}R_{jl}, \tag{9.5}$$

where ω denote the IO-coefficient from sector j to i and n the sectoral bilateral trade flow from country l to k. Thus, both changes in the weights and changes in the different R&D-capital stocks affect the knowledge-stock construct. According to equation (9.5) the spillover stocks from different sectors or countries are complements. This is based on the idea that the more ideas are available the higher productivity is.

We introduce an additional assumption in the construction of the R&D spillovers. So far we assumed that (trade-weighted) R&D stocks are complements. We, however, argue that the sources of the (sectoral and regional) R&D stocks are imperfect substitutes.[13] To be more specific we assume that the relative likeliness that the R&D stock from sector j in country l is valuable for sector i in country k is inversely related to the size (of value added) of sector j. This assumption is motivated by the notion that if you import from a large sector with consequently a relatively large R&D stock, a relatively small *share* of that sector's R&D stock will be embodied in these imports.[14] The R&D stocks can than be written as:

$$\begin{aligned} R_{ik}^{ID} &= \sum_{j\neq i}^{I} \omega_{jik}\, \eta_{jik}\, R_{jk} \qquad \eta_{jik} \equiv \frac{R_{ik}^{ID}}{Y_{jk}} \\ R_{ik}^{F} &= \sum_{l\neq k}^{K} \sum_{j}^{I} n_{jilk}\, \mu_{jilk}\, R_{jl} \qquad \mu_{jikl} \equiv \frac{R_{ik}^{F}}{Y_{jl}}, \end{aligned} \tag{9.6}$$

where η and μ are the size correction factors. Assuming that only the R&D stocks are time dependent, time differentiation of equation (9.6) leads to

$$\begin{aligned} \hat{R}_{ik}^{ID} &= \sum_{j\neq i}^{I} \frac{\omega_{jik}\, \eta_{jik} R_{jk}}{R_{ik}^{ID}}\, \hat{R}_{jk} \\ \hat{R}_{ik}^{F} &= \sum_{l\neq k}^{K} \sum_{j}^{I} \frac{n_{jilk}\, \mu_{jilk} R_{jl}}{R_{ik}^{F}}\, \hat{R}_{jl}\ , \end{aligned} \tag{9.7}$$

Moreover we assume that R&D stocks do not depreciate. Here we follow the Terlecky approach (1974).[15] That is, the growth of the R&D knowledge stock can be approximated by the R&D expenditures if is assumed that the depreciation of the stock is zero. As a result equation (9.7) becomes (substitute also for the ηs and μs)

$$\hat{R}_{ik}^{ID} = \sum_{j \neq i}^{I} \omega_{jik} \frac{RD_{jk}}{Y_{jk}} \qquad \hat{R}_{ik}^{F} = \sum_{l \neq k}^{K} \sum_{j}^{I} n_{jikl} \frac{RD_{jk}}{Y_{jk}} \ . \tag{9.8}$$

The construct in equation (9.8) exhibits several desirable characteristics. First, it does not suffer from an aggregation bias, as equation (9.4) does. The latter construct is very sensitive to statistical aggregation of countries (see Lichtenberg and Pottelsberghe de la Potterie, 1998, and Jacobs et al., 2002).[16] We have solved this problem by the assumption that spillovers are related to the size of the country. This approach is largely insensitive to aggregation as it avoids weighting the growth rates of large countries or large delivering sectors heavily. The adjustment of the weights can thus be interpreted as that we allow the γs to be specific for every sector and region.

Second, a change in the knowledge stocks over time, now approximated by the R&D intensities that vary over time, is captured in (9.8). Third, a change in the weight matrix affects the spillover construct. Now this is only the case if R&D investments are positive whereas in a specification with R&D stocks, integration-induced changes in the weights would affect the R&D construct directly.[17] Related is that by using R&D intensities (equation (9.8)) instead of an equation based on uncorrected weighted levels (equation (9.7) without country-sector specific ηs and μs) integrating a country with an average R&D intensity in the global economy has no effect on the R&D construct.

Fourth, we introduce another effect of global integration that is easily clarified by discussing the weighting coefficients, ω and n. We use the following definitions:

$$\omega_{ijk} = \frac{U_{ijk}^{D}}{Y_{ik}^{G}} \qquad n_{ijkl} = \frac{U_{ijk}^{M}}{Y_{ik}^{G}} \frac{M_{jkl}}{\sum_{l} M_{jkl}} \ , \tag{9.9}$$

where U indicates intermediate-input use (superscripts D and M stand for domestic and imported) and Y^G denotes gross production. Hence, integrating a formerly isolated country, with an average R&D intensity, in the global economy will affect the knowledge spillover *if* the import quote, approximated by the imported use over gross production, goes up. This interpretation closely follows a returns to variety production function (see de Groot and Nahuis, 1998). Hence, if the intermediate inputs of an economy are useful, spillovers increase.

We write our estimating equation (equation (9.4)) as:

$$\begin{aligned}\hat{F}_{ikt} &= c + \beta^{DD}\frac{RD_{ikt}}{Y_{ikt}} + \beta^{ID}\sum_{j \neq i}^{I} \omega_{jikt}\frac{RD_{jkt}}{Y_{jkt}} \\ &+ \beta^{F}\sum_{l \neq k}^{K}\sum_{j}^{I} n_{jilkt}\frac{RD_{jkt}}{Y_{jkt}} + \nu_{ikt} \, .\end{aligned} \tag{9.10}$$

The results we report on in the next section are based on this expression. Note that the explanatory variables in (9.10) are not literally growth rates as we substitute (9.8) in (9.4). We replace the γs by βs to stress the relation between the parameters we estimate and those that will be implemented in the AGE model. The γs we implement are thus bi-laterally country and sector specific (as is required by our assumption motivated above; see Appendix C for further details).

3. DATA AND ESTIMATION

In this section we present the data that are used in the estimation procedure and the model. First, the dependent variable, TFP, is constructed. Second, we describe R&D intensities for OECD and non-OECD regions. Finally, we present estimation results for the specification discussed above.

3.1 TFP growth rates

Explaining differences in technological efficiency is our main interest. Our preferred indicator of technological efficiency is Total Factor Productivity (further TFP) as it measures the efficiency of the combined capital and labour inputs. This section shows how we measure TFP growth.

We calculate the growth rate of TFP ($\hat{F}$) for industry i in country k at time t by a superlative index that is consistent with a translog function (see Diewert, 1976):

$$\hat{F}_{ikt} = \ln\left(\frac{Y_{ikt}}{Y_{ikt-1}}\right) - 0.5(\alpha_{ikt} + \alpha_{ikt-1})\ln\left(\frac{L_{ikt}}{L_{ikt-1}}\right) - (1 - 0.5(\alpha_{ikt} + \alpha_{ikt-1}))\ln\left(\frac{Q_{ikt}}{Q_{ikt-1}}\right) . \qquad (9.11)$$

Value added, Y is in PPP-converted constant US dollars. Employment, L, is the number of workers employed, Q is the capital stock estimated by the OECD and α is the wage-income share (see OECD, 1999a).

Table 9.1 presents the mean annual TFP growth rates for the different regions, from 1973 to the early 1990s. The numbers presented in the table are generated with OECD (1999a) data, using equation (9.11) as a starting point. We have adjusted employment for hours worked.[18]

Across countries as well as across sectors there is considerable variation in the growth rates. Some results are worth emphasizing. Europe's current strong position in energy-intensive goods (including, for example, the chemical sector) is the result of a high growth rate throughout the period. Remarkable is the low productivity growth in consumer goods in Japan. The R&D-intensive capital goods sector is the most dynamic sector where the productivity leader – the US – has the lowest average growth rate.

Table 9.1 Mean annual TFP growth rates (adjusted for hours worked)

	Agriculture	Raw materials	Energy-intensive goods	Consumer goods	Capital goods	Total services
US	1.91	-0.94	0.39	0.96	1.50	0.13
Japan	-3.79	-0.65	-0.82	-0.40	4.19	-0.19
R-OECD	1.66	-2.38	1.13	0.79	2.24	0.35
W-Europe	1.01	1.09	2.75	1.46	2.80	1.12

Note: [a] For the US, Japan and Rest-OECD the data refer to 1971–93. For Western Europe to 1971–91.
Source: OECD (1999b), Maddison (1991) and own calculations.

The Services sector has experienced hardly any productivity growth, except for Western Europe. The relative backwardness of the Japanese agriculture sector arose due to low productivity growth in the last three decades whereas productivity growth in US agriculture is relatively high. The results for raw materials have to be taken with a grain of salt, given the fact that this sector has been affected considerably by the oil crisis.

In general Europe's growth rate exceeds the US' as a consequence of catching up. The low Japanese TFP growth rates, despite substantial economic growth, reflect the considerable capital deepening and increases in labour participation that are observed in the last decades throughout Asia (see, for example, Young, 1995, and Kim and Lau, 1994).

3.2 R&D intensities

The size and importance of R&D spillovers between countries and industries depends to a large extent on the knowledge stocks in the different sectors and countries. First, this section describes the observed R&D intensities. These are based on data from OECD (1999a) and UNESCO (1998, 1999). The former provides business-enterprise data for the OECD at a sectoral level, the latter provides business-enterprise data for the non-OECD economies at a macroeconomic level. Second, we discuss the construction of the sectoral and regional business-enterprise R&D intensities in WorldScan.

OECD regions

The ANBERD data base of the OECD (1999a) provides the value of R&D expenditures for business enterprises of 15 OECD countries from 1973 to 1997 at a sectoral level according to the ISIC2 classification. The data are highly disaggregated for the manufacturing sectors but not for services. Moreover, for most countries no data for agriculture and mining (raw materials) are included. The ANBERD data base contains a residual (total, minus R&D in manufacturing and services) which has to be split up between agriculture and mining.[19]

We combined the ANBERD data with the ISDB data (OECD, 1999b) to derive R&D intensities per sector for the various countries. The latter database provides value added data at a sectoral level. This enables us to derive R&D intensities per sector and country. Table 9.2 reports these for the four OECD regions in WorldScan. In order to derive the sectoral business enterprise R&D intensities for the OECD regions in WorldScan, we simply aggregate the country data to WorldScan sectors and regions. We assume that the underlying country data (see Table A.1) are representative for the relevant WorldScan regions.[20]

In general, the R&D intensities in the raw materials and agriculture sectors are higher than in services, but lower than in manufacturing. The variation within manufacturing is interesting. The R&D intensities in energy-intensive goods and capital goods are very high, while they are relatively low in consumer goods. The latter consists of sub-sectors like wood, food and tobacco, textiles and paper which are R&D extensive sectors. The sector energy-intensive goods is R&D intensive because the sub sector chemicals, rubbers and plastics is included. The R&D intensity of other sub sectors like stone and clay and basic metals is lower. The capital goods sector consists only of fabricated metal products, which is very R&D-intensive.

Table 9.2 Sectoral R&D intensities in WorldScan for the OECD, as ratio of sectoral value added (1990)

Sectoral R&D	Agri-culture	Raw materials	Consu-mer goods	Energy -int. goods	Capital goods	Services	Average
Western Europe	0.62	0.96	0.59	4.49	9.39	0.23	1.81
United States	0.53	0.53	1.11	5.23	15.22	0.53	2.21
Japan	0.10	2.65	1.16	8.10	10.64	0.12	2.25
Pacific OECD	0.18	0.46	0.61	2.42	7.07	0.41	1.00
Average OECD	0.45	0.68	0.92	5.34	11.84	0.36	2.05

Source: OECD (1999a, 1999b) and own calculations.

If we compare the regions, we see that the United States and Japan carry out most of the R&D while Pacific OECD is lagging behind. The United States carries out relatively high levels of R&D in capital goods, consumer goods and services, while Japan is active in energy-intensive goods and raw materials. Western Europe carries out a lot of R&D in agriculture.

Non-OECD regions

UNESCO (1999) provides, for about 100 countries, the expenditures on R&D as ratio of Gross National Product for several years in the 1980s and 1990s. For most countries data are limited to a few years. The coverage, however, is wide. The R&D intensities vary widely among the countries. In general these intensities are much lower for developing countries than for the industrial countries. Table 9.3 presents the results for the non-OECD WorldScan regions. R&D in the most developed region, South-East Asia, is the highest.[21]

Table 9.3 R&D intensities for the non-OECD regions in 1995

R&D intensity	Total R&D[b]	Share BE[a]	BE R&D[b]
Eastern Europe	0.90	47.1	0.42
Former Soviet Union	0.73	67.1	0.49
Latin America	0.55	29.5	0.16
Middle East	0.76	41.6	0.32
Sub-Saharan Africa	0.61	52.7	0.32
China	0.61	31.9	0.19
South-East Asia	1.33	70.0	0.93
South Asia & Rest	0.69	26.5	0.18

Notes: [a] BE = business-enterprise R&D (as a share of total R&D).
[b] as ratio of GNP.
Source: UNESCO (1998, 1999).

For some regions the coverage is limited to a few countries, such as Sub-Saharan Africa (only South Africa) and Middle East (only Turkey and Israel). The coverage for Former Soviet Union and Eastern Europe, China, South Asia & Rest and South-East Asia is fairly good. The business enterprise R&D intensities vary more widely than those of total R&D. The numbers on the share of business enterprise R&D reinforce the differences; see for example the effects on China and South-East Asia.

3.3 Empirical findings

This section presents the main empirical findings. The model in equation (9.10) is estimated for all sectors WorldScan distinguishes. Appendix B presents robustness analysis and results for the manufacturing sectors only. The results presented in this section will be made operational in the AGE model in the next section.

One remark should be made beforehand. Our regression analysis has only two aims. First, we want to establish that the relations, found in the literature,[22] can also be traced at the aggregation level of WorldScan. Second, the estimates should provide parameters for the AGE model we employ in the next section.

The aggregate model

Table 9.4 presents the regression results based on equation (9.10).

Table 9.4 OLS results for equation (9.10). Dependent variable is $T\hat{F}P$.†

Variable	(I) Direct effect	(II) Direct + indirect effect	(III) Domestic and Total
DD	$.216^{***}$	$.205^{***}$	$.167^{**}$
	[.069]	[.069]	[.074]
ID		2.112^{**}	2.636^{**}
		[.966]	[1.041]
TF			0.618
			[.457]
R^2 (adjusted)	0.02	0.03	0.03
N	432	432	432

Notes: † The sample period is 1973–91. The sample consists of 6 sectors and 4 regions. All regressions include a constant. The explanatory variables are lagged by one year. Standard errors are given in parentheses under the estimates.
, and * denote statistical significance at the 5 per cent level, and the 1 per cent level, respectively.

First, we include own R&D stocks. We find a significant rate of return for the own within-sector R&D stock. Inclusion of the indirect domestic R&D stock in column (II) supports the hypotheses that within-region R&D spillovers exist. The estimated coefficient for the indirect effect is relatively high, because we use weighting matrices for which the columns do not add up to unity. Inclusion of the foreign spillover variable in column (III) yields an estimate for our foreign R&D construct of 0.6. The estimate is not very precise, however. The inclusion does not substantially affect the coefficients for the domestic variables. This regression is our major input for the modelling exercise in the next section.

How do these findings match with the literature? The initiating contribution on international spillovers is the paper by Coe and Helpman (1995), further CH. They find substantial technological spillovers among OECD countries. The elasticity of the level of TFP with respect to foreign R&D embodied in traded goods is about 6 per cent. Park (1995) examines also country-level data without an industry dimension. Labour-productivity growth is explained by domestic R&D and foreign R&D weighted by technological distance. The elasticity of weighted foreign R&D is 17–18 per cent compared to 11 per cent for domestic R&D.

Keller (1997) carries out a similar exercise to ours, be it on a more disaggregated level. R&D in the 'same' sector abroad turns out to have an equally strong effect on TFP as R&D carried out by the sector itself. Verspagen (1997) estimates production functions and constructs the R&D spillover stock by using weights differently from ours. He uses R&D stocks in a similar vein to CH. In the most comparable estimate to ours, he finds an own-R&D elasticity of 9 to 10 per cent. The indirect domestic elasticity varies from 3 to 6 per cent, and the foreign from 5 to 7.5 per cent. To transform the estimated coefficient of *ID*, presented in the table, to a number comparable to one where a weighting scheme with columns summing up to unity is used, one should divide the estimated coefficient roughly by a factor of 5.[23] This implies for regression (III) in Table 9.4 an elasticity of 30 per cent, which is considerably higher than for example Verspagen's. Sakurai et al. (1997) use comparable data and estimate rates of return; they find an own-R&D elasticity of 13 to 17 per cent. Our coefficient for foreign R&D seems rather low (note that we pre-multiplied the R&D intensities twice with a weighting coefficient with a sum less than unity).

4. WORLDSCAN: A GLOBAL APPLIED GENERAL-EQUILIBRIUM MODEL

WorldScan has been developed to construct scenarios. WorldScan relies on the neoclassical theories of growth and international trade.[24] The standard neo-classical theory of growth distinguishes three factors to explain changes in production: physical capital, labour and technology. WorldScan augments the simple growth model in three ways. First, WorldScan allows overall technology to differ across countries. Second, the model distinguishes two types of labour: high-skilled and low-skilled labour. Sectors differ according to the intensity with which they use high-skilled and low-skilled labour. Countries can raise per capita growth by schooling and training the labour force. Third, in developing countries part of the labour force works in a low-productivity, informal sector. In this sector workers do not have access to capital and technology. Reallocation of labour from the low-productivity sector to the high-productivity sectors enables countries to raise per capita growth as well. In principle – consistent with the neo-classical growth theory – all these three factors affect the performance of a region only temporarily. Catching-up of technology, training of low-skilled workers and reallocating labour to the high-productivity sector do not raise the growth rate indefinitely.

The model applied in this chapter differs considerably from that in Chapter 8. A comprehensive overview of the Chapter 8 version is provided in Geurts et al. (1995) whereas this chapter's version is presented in detail in CPB (1999). Here we highlight only a few differences. The model in this chapter has less rigidities on labour and goods markets (unemployment is exogenous and there is no inventory effect on prices). The input-output table of the model is completely filled in accordance with the data, whereas in the Chapter 8 version, the application of factor proportions-theory was guiding the input-output structure (see Nahuis, 1999, for an elaboration). Where technology in Chapter 8 was governed by an (ad hoc) technological catch-up mechanism, the version on the *basis* of this chapter has exogenous technological change.

The simulations in Section 5 are variations on the so-called globalization scenario.[25] The idea behind the scenario is that when developing countries grow fast or start to grow rapidly, the linkages between the OECD and the non-OECD countries intensify. Fast development outside the OECD area and liberalization of capital, goods and service markets produce closer economic integration of rich and poor countries. More generally, the scenario extrapolates and probably exaggerates the current globalization tendencies.

The globalization scenario is optimistic about future economic progress in both developed and developing regions. In this scenario many poor countries catch up, though not completely, with rich countries. Non-OECD countries grow at a per-capita rate of about 4 per cent. Only few countries have been able to maintain such a growth rate for two decades or more.

In the scenario, trade liberalization is not confined to trade blocs, but applies globally. The OECD countries open up their markets further. Whereas barriers to trade in manufacturing goods are already low, agriculture is still heavily protected in the globalization scenario. Table 9.5 provides an overview of the average import and export taxes in the OECD and non-OECD in 1995.

Within manufacturing, the import tariffs are the highest for consumer goods. For energy-intensive goods and capital goods which are often used as intermediate goods the import tariffs are lower. The non-OECD regions often levy higher tariffs than the OECD regions. The tariffs in the service sectors are very low, but the non-tariff barriers are very high. The OECD countries subsidize their agricultural and food products (food is a substantial part of consumer goods). For this reason the export taxes are negative.

Even though the globalization scenario is perhaps not the most plausible one, we take it as point of departure. The reason is that it stresses that linkages between developed and developing regions can become stronger and spillovers between these regions can become larger.

Table 9.5 Trade taxes in the OECD and the non-OECD in 1995

Sectors	Agri-culture	Raw materials	Energy -int. goods	Consumer goods	Capital goods	Trade and trans-port	Services
Average import tariffs (%)							
OECD	32.0	0.4	2.9	11.0	2.6	1.1	0.3
Non-OECD	18.6	5.1	11.8	19.7	12.1	0.2	0.5
Average export tariffs (%)							
OECD	-2.9	1.1	0.6	-3.9	0.3	3.7	2.4
Non-OECD	3.8	2.3	-0.6	2.6	0.1	1.1	1.2

Note: A minus sign implies a subsidy.
Source: McDougall et al. (1998) and own calculations.

4.1 Incorporation of R&D in the model

WorldScan has been calibrated on the GTAP data base, Version 4 (McDougall et al., 1998). From this data set we not only derive the demand, production and trade patterns, but also the labour and capital intensity of the different sectors. The incorporation of R&D affects the model and the data. To start with the latter, our base-year data derived from the GTAP database do not include expenditures on R&D. We assume that these are implicitly incorporated in the intermediate deliveries on services. Therefore, we subtract the expenditures on R&D from the GTAP data on intermediate deliveries on services. As described before the R&D data are derived from the OECD (1999a) and UNESCO (1998) data for the base year 1995. We also subtract R&D expenditures from the value of production. Based on the modified GTAP data we calibrate the production function. Then we construct a new producer price as the unit cost price plus a mark up which covers the R&D expenditures. As a result, the volume of production times the new producer price is equal to the production value in the original GTAP data. Total demand for services now consists of intermediate demand, investment demand, final consumption demand and R&D demand. The total value of the demand for services is still the same as in GTAP.

Reiterating for Section 2, TFP is a function of R&D stocks and the R&D-spillover stocks

$$F_{ikt} = A_{ikt}\left(R_{ikt}^{DD}\right)^{\gamma^{DD}}\left(R_{ikt}^{ID}\right)^{\gamma^{ID}}\left(R_{ikt}^{F}\right)^{\gamma^{F}}, \qquad (9.12)$$

γs are the adjusted sector and country coefficients estimated in Section 3. This adjustment is necessary to translate the estimated equation in the equation above. Details are provided in Appendix C. The constructs for own R&D stocks, indirect and foreign R&D stocks were defined in equation (9.5).

The sectoral R&D stocks in period t equal those in period t-1 – corrected for deprecation – plus the R&D expenditures. The depreciation rate δ is set at 5 per cent for the R&D stock in all sectors and regions.[26] The R&D expenditures are by assumption a constant fraction, $\overline{RI}$, of sectoral value added in period t-1, thus

$$R_t = \overline{RI}\,Y_{t-1} + (1-\delta)R_{t-1}\,. \qquad (9.13)$$

We also use this equation to construct the R&D stock for the base year assuming that the ratio of the R&D stock to value added is constant. The

R&D stock equals

$$\frac{R_0}{Y_0} = \frac{\bar{RI}}{\delta + g_Y}, \tag{9.14}$$

in which g_Y represents the growth rate of value added. In the base year we calibrate A by inverting equation (9.12) where we substitute the values of the R&D stocks. The value of F follows from calibrating the production function.

In the scenario period, TFP grows due to an exogenous increase in A and an endogenous increase in the R&D stocks. In the baseline without R&D we have imposed an exogenous increase in sectoral and regional TFP in the model such that the model produces the characteristics of the globalization scenario. In the baseline simulations including R&D we have assumed that the total increase in TFP was similar to that in the baseline without R&D. As a result, the exogenous increase in A is much lower in the simulations with R&D than without R&D. We follow this method to make the baselines comparable to each other. The effects of trade liberalization are then also comparable.[27]

5. SIMULATION RESULTS

This section presents the effects of trade liberalization in case R&D is introduced in WorldScan. We distinguish the effects of trade liberalization in the presence of own R&D efforts, of sectoral spillovers and of international R&D spillovers. These effects are measured by comparing the results for two simulations: a baseline simulation without trade policy and a policy variant consisting of trade liberalization. First, we present the results of introducing R&D on GDP growth for the various regions in the baseline simulation.

Second, we turn to the macroeconomic effects of trade liberalization and the role of R&D (spillovers). Third, we discuss the sectoral effects for some regions.

Based on the results we carry out some sensitivity analysis. We analyse the effects from an increase in R&D efforts in the non-OECD countries, which is to be expected if these regions become more wealthy.[28]

5.1 Growth accounting

The incorporation of R&D and spillovers in our baseline simulation has a significant effect on GDP growth in the model. While thus far a substantial part of GDP growth was explained by TFP growth (CPB, 1999), the contribution of exogenous TFP growth is declined in favour of growth in R&D and R&D spillovers. Table 9.6 shows the factors that contribute to GDP growth in the various regions.

CPB (1999) explains that a substantial part of GDP growth in the non-OECD regions can be attributed to the growth in employment. This is caused by population growth, schooling and labour reallocation from the low-productivity sectors to the high-productivity sectors. On average capital accumulation contributes about 40 per cent to GDP growth. The rest can be attributed to R&D and TFP. This is our main interest here.

According to Table 9.6 R&D explains a part of GDP growth which was attributed to TFP before. Own R&D only contributes significantly to GDP growth in the OECD and South-East Asia, the regions that perform nearly all R&D in the world. The relevance of the sectoral and international spillovers varies per region. Below we will discuss these issues at greater length. Table 9.6 shows that for most regions the spillovers contribute more to GDP growth than own sectoral R&D efforts. This is not surprising. In particular, the sectoral spillovers are mainly driven by those goods which are relatively important as intermediate goods such as capital goods and energy-intensive goods. These sectors are also relatively R&D intensive. This implies that the contribution of sectoral spillovers to GDP growth is larger than the contribution of own R&D.

Table 9.6(a) Growth accounting

Country	Annual contributions of the productive factors (1995–2020)					
	Western Europe	United States	Japan	Pacific OECD	Eastern Europe	Former Soviet Union
Employment	−0.1	0.4	−0.2	0.4	0.0	0.2
Capital accumulation	0.8	1.0	1.1	1.0	1.3	2.1
Own R&D (R^{DD})	0.1	0.2	0.1	0.1	0.0	0.0
Sectoral R&D spillovers (R^{ID})	0.1	0.2	0.1	−0.1	0.1	0.1
International R&D spillovers (R^{F})	0.1	0.1	0.0	0.2	0.2	0.1
Total factor productivity (A)	1.4	0.8	1.2	0.6	2.9	2.9
Gross domestic product	2.4	2.7	2.4	2.1	4.5	5.5

Source: WorldScan simulations.

5.2 Trade liberalization and GDP effects

The growth-accounting analysis learns that a part of TFP growth can be explained by R&D. R&D growth thus raises GDP growth. This result is also confirmed in our analysis of trade liberalization. Without R&D in the model, the effects of trade liberalization on GDP are in general modest. We want to examine whether this is also the case if R&D is included in WorldScan. We carry out a trade-liberalization exercise in four different cases. These cases are discriminated by the fact that:

- TFP is not affected by R&D.
- TFP is only affected by own R&D expenditures.
- TFP is affected by own R&D and sectoral spillovers.
- TFP is affected by own R&D and sectoral and international spillovers.

Table 9.6(b) Growth accounting

Country	Middle East & North Africa	Sub-Saharan Africa	Latin America	China	South-East Asia	South Asia & Rest
Employment	1.6	2.7	1.3	0.7	1.4	1.8
Capital accumulation	3.1	2.0	2.5	4.2	3.1	2.6
Own R&D (R^{DD})	0.0	0.0	0.0	0.0	0.1	0.0
Sectoral R&D spillovers (R^{ID})	0.1	0.1	0.1	0.2	0.5	0.1
International R&D spillovers (R^{F})	0.1	0.1	0.1	0.2	0.2	0.1
Total factor productivity (*A*)	0.8	0.2	1.0	1.9	1.2	1.3
Gross domestic product	5.7	5.1	4.9	7.2	6.4	5.9

Source: WorldScan simulations.

The first simulation assumes no link between R&D and TFP. We assume that all regions agree to abolish all their tariffs and export subsidies between 2000 and 2020. In the agriculture and raw materials sectors the import tariffs and export subsidies are reduced by only 50 per cent, because of the high initial rates of protection in these sectors. The results are similar to those in Lejour and Tang (2000). The effects on GDP in the OECD are modest, but the Asian regions gain substantially in 2020, the end of the simulation period. Also the GDP gains in Latin America are large.[29] The first column in Table 9.7 presents these results.

The second simulation assumes that increases in the sectoral R&D stock raise the TFP level in that sector. This simulation does not take into account the sectoral and international spillovers on TFP. Column (2) shows the additional GDP effects of trade liberalization on GDP due to own R&D expenditures. These extra effects are modest, except for Western Europe, Japan and South-East Asia. These regions specialize in capital goods and energy-intensive goods. Trade liberalization positively affects growth in these sectors and thereby stimulates the R&D efforts.

Column (3) shows the additional GDP effects of trade liberalization due to the sectoral spillovers. These effects vary widely. In South-East Asia the sectoral R&D spillovers increase the GDP effects of trade liberalization with 6 percentage-points. In Sub-Saharan Africa and China however, the sectoral spillovers have a small negative effect on GDP. The results vary by region because of the regional differences in the development of the R&D-intensive sectors. From Table 9.2 we know that the capital goods and energy-intensive goods sectors are R&D intensive. In regions that do not specialize in these sectors, the R&D-intensive sectors become relatively less important during the process of trade liberalization. Then, the average R&D content of the intermediate goods produced in the own region decreases. Examples are Sub-Saharan Africa and China. In other regions the R&D-intensive sectors expand relatively quickly. As a consequence, the average R&D content of the intermediate goods increases. This explains the sectoral spillovers in Western Europe, Japan and South-East Asia. Thus, the importance of sectoral spillovers depends on the specialization pattern. Regions can specialize in R&D-intensive or R&D-extensive sectors. We will discuss this issue in greater detail below.

Table 9.7 Cumulative GDP effects of trade liberalization in 2020

Region	No R&D (1)	Own R&D (2)	Sectoral R&D spillovers (3)	International R&D spillovers (4)	Relative GDP increase due to R&D (5)
United States	1.5	0.0	0.3	0.4	53.2
Western Europe	1.7	0.5	2.3	0.1	169.3
Japan	2.3	1.0	7.4	0.2	372.1
Pacific OECD	3.8	0.2	0.5	0.2	24.1
Eastern Europe	5.0	0.2	1.0	0.8	40.2
Former Soviet Union	1.6	0.1	0.2	0.2	36.2
Latin America	9.5	0.3	0.7	0.4	14.6
Middle East & N. Africa	4.8	0.3	0.4	2.1	58.8
Sub-Saharan Africa	5.0	0.1	-0.3	0.7	10.2
China	15.0	0.1	-0.5	0.9	3.4
South-East Asia	14.9	1.4	6.0	1.5	59.6
South Asia & Rest	15.9	0.2	0.4	0.6	7.0

Source: WorldScan simulations. Some simulation results without depreciation of R&D stocks are presented in Appendix D.

The international R&D spillovers further raise the GDP effects of trade liberalization, as can be seen in column (4) of Table 9.7. Its importance differs per region. In general, international R&D spillovers are more important for the non-OECD regions than for the OECD regions. Non-OECD regions import relatively much from the OECD, whose products are relatively R&D intensive, see also Table(s) 9.2 (and 9.3). An extreme example is the Middle East. This region imports more capital goods and energy-intensive goods from the OECD due to trade liberalization. As a result the international spillovers are high.

Column (5) shows the increase in the GDP effects of trade liberalization with R&D in the model relative to the GDP effects of trade liberalization without R&D. On average the GDP effects are raised significantly (due to R&D-based technology). China, Sub-Saharan Africa and South Asia and Rest are exceptions, however.[30]

Sectoral and international spillovers

Above we have seen that the large variety in GDP effects of trade liberalization due to sectoral spillovers depends on the development of the R&D-intensive sectors. The energy-intensive goods and capital goods sectors are very important in this respect for two reasons. First, these sectors are very R&D intensive. Second, these goods are intensively used as intermediate goods. Table 9.8 presents some indicators of the development of these sectors and their effects on regional R&D stocks in the process of trade liberalization.

Western Europe, Japan, Eastern Europe and South-East Asia specialize in R&D-intensive sectors (see, column (1)). These sectors are also high-skilled labour intensive, which largely explains specialization in these sectors by the former three regions (which are relatively high-skilled abundant). In these regions, the share of R&D-intensive sectors in value added rises. This enhances the growth of the R&D stocks in these sectors and positively affects the regional R&D stocks. The last two columns in Table 9.8 show a high positive correlation between the changes in the R&D stocks of the R&D-intensive sectors and the regional R&D stock. So, the sectoral spillovers are very high in regions which tend to specialize in the production of R&D-intensive goods. The United States and Pacific OECD specialize in agriculture which is R&D extensive. As a consequence, their sectoral spillovers are modest.[31]

Table 9.8 Development of R&D-intensive sectors due to trade liberalisation in 2020

Region	Absolute change in share of R&D-intensive sectors in value added[a] (1)	Relative change in R&D stocks of R&D-intensive sectors (2)	Relative change in total R&D stocks (3)
United States	-0.7	-1.5	-0.9
Western Europe	2.3	10.4	8.6
Japan	3.9	17.9	16.2
Pacific OECD	-1.0	-0.7	1.1
Eastern Europe	0.7	8.9	8.4
Former Soviet Union	-1.0	-1.6	-0.7
Latin America	-0.6	2.9	4.3
Middle East & N. Africa	-1.3	-5.3	-3.6
Sub-Saharan Africa	-3.3	-13.4	-9.2
China	-7.2	-9.9	-5.6
South-East Asia	3.9	30.6	28.3
South Asia & Rest	-1.4	2.9	6.3

Note: [a]The R&D-intensive sectors are the energy-intensive goods and capital goods sectors.
Source: WorldScan simulations.

The negative sectoral spillovers in China and Sub-Saharan Africa in Table 9.7 can be explained in a similar way. These regions specialize in consumer goods and agriculture, respectively, at the expense of R&D-intensive goods. So their regional R&D stocks decrease if trade liberalization takes place. The sectoral spillovers for trade liberalization are thus negative for these regions.

The size of the international spillovers can analogously be explained by the R&D content of the imports. These spillovers depend on the structure of the imports. Here the origin of imports is important as well as the sectoral composition. Table 9.9 illustrates this.

The importance of international R&D spillovers is determined by the R&D content of the imports. Column (3) in Table 9.9 shows the relative increase in the R&D content of the imports. It is very large for the Middle East, which explains the large international spillovers on GDP, see Table 9.7. The large increases in the R&D content of the imports in the United States and South Asia also lead to relatively high international R&D spillovers.[32]

Table 9.9 R&D content of imports due to trade liberalisation in 2020

Region	Absolute change in share of OECD imports in total imports (1)	Relative change in R&D content of import of R&D int. sectors (2)	Relative change in R&D content of total imports (3)
United States	-5.4	24.1	22.2
Western Europe	-4.2	12.6	11.0
Japan	0.9	13.3	13.4
Pacific OECD	-7.7	5.3	4.8
Eastern Europe	-2.8	15.7	13.3
Former Soviet Union	-5.1	12.1	11.0
Latin America	-12.6	5.1	4.2
Middle East & N.A.	7.8	51.5	46.2
Sub-Saharan Africa	-9.6	10.7	9.9
China	-3.5	8.3	9.5
South-East Asia	0.4	8.0	7.6
South Asia & Rest	-4.3	18.2	16.8

Source: WorldScan simulations.

The changes in the R&D content of the imports are affected by the changes in the regional and sectoral structure of the imports. The columns (1) and (2) present two indicators for these changes. Table 9.9 shows that most regions import less from the OECD; the OECD regions that have the highest R&D stocks. The reason is that trade liberalization affects the relative consumer prices. The relative price of goods from the non-OECD regions tends to become lower due to the elimination of tariffs. Only Japan, the Middle East and South-East Asia import relatively more from the OECD after trade liberalization. This has a positive effect on the R&D content of the imports for these regions.

The changes in sectoral structure of these imports are very important. All regions import relatively more R&D because they import relatively more R&D-intensive goods. Trade liberaliaation stimulates, in particular, trade in manufacturing products. The reason is that the fall in trade barriers in these sectors is larger than in services and raw materials. In particular the United States and the Middle East import more of these goods, which leads to a considerable rise in the R&D content of the imports. So, although the sectoral spillovers in the United States are low, because it specializes in agriculture, the international spillovers are high due to the increased imports of energy-intensive and capital goods. Table 9.9 thus shows that the international spillovers are high if regions import relatively more R&D-intensive goods and if the imports originate from the OECD.

Above we have analysed the contribution of R&D and its spillovers on the effects of trade liberalization. R&D magnifies the positive effects on GDP if the R&D content of the intermediate goods is high. This can be achieved in two ways. The first is that regions carry out a lot of R&D themselves. The second is that they import relatively many R&D-intensive goods. The analysis shows a trade-off between specializing in R&D-intensive goods and importing these goods. Western Europe, Japan, Eastern Europe, Latin America and South-East Asia produce relatively many R&D-intensive goods. On the other hand, their R&D-import content is low. These regions thus have high sectoral spillovers compared to the international R&D spillovers (see Table 9.7).

The United States, the Middle East, Sub-Saharan Africa, China and South Asia & Rest import relatively many R&D-intensive goods. The contribution of R&D to the GDP effects of trade liberalization are mainly through international spillovers in these regions whereas they experience low or even negative effects related to the sectoral spillovers.

Wage inequality

In spite of the relatively large GDP effects of trade liberalization, the relative wages of low-skilled workers are hardly affected in the OECD. Table 9.10 shows the effects of trade liberalization on the ratio of wages of high-skilled workers to those of low-skilled workers.

Table 9.10 Impact of trade liberalization on relative wages in 2020 relative wage of high- to low-skilled workers

Region	No R&D (1)	Own R&D (2)	Sectoral R&D spillovers (3)	International R&D spillovers (4)	Total (5)
United States	-0.7	-0.1	0.0	0.0	-0.8
Western Europe	1.2	0.1	0.0	0.0	1.3
Japan	2.2	0.1	0.1	0.0	2.3
Pacific OECD	-1.4	0.0	0.0	0.0	-1.4

Note: Numbers are absolute changes.
Source: WorldScan simulations.

The columns in Table 9.10 distinguish the four different cases with and without R&D spillovers. The wage ratio of the four OECD regions is about 1.61. The numbers in the table refer to absolute changes in this ratio. The effects of trade liberalization on the labour markets are very modest in the simulation without R&D spillovers. In Western Europe and Japan the skill premium rises with 1 to 2 percentage-points. In the other two regions the skill premium decreases slightly, because those regions specialize in agriculture. Lejour and Tang (2000) derive a similar result. The introduction of R&D and R&D spillovers in TFP exerts a mild upward pressure on the skill premium in Western Europe and Japan. The reason for this small effect is (i) the low sensitivity of TFP for 'own'-sector R&D and (ii) the fact that high-skill/R&D-intensive sectors generate considerable spillovers within the domestic economy towards low-skilled intensive sectors. The effect for the US is explained by the fact that this region does not specialise in high-skill/R&D-intensive sectors. From this analysis we conclude that R&D spillovers do not significantly affect the position of low-skilled workers.

Comparing the results for wage inequality here with those in the previous chapter reveals some remarkable differences. The relatively modest numbers presented in Table 9.10 compared to those in Table 8.5 (see Chapter 8) can be explained by differences in the data, the model and the set-up of the experiment. First, this version of WorldScan is calibrated on the GTAP data base, whereas the version used in Chapter 8 was calibrated on a data set constructed by CPB, see Geurts et al. (1995). Second, the translation of the data to the model differs importantly. The Chapter 8 version had a relatively empty IO-table which implied that considerable data manipulation was necessary to replicate aggregate data. This rearranging of the data was done with factor-proportion theory in mind; that is, the production structure motivated the assignment of output to a certain sector. In the GTAP-based version the data are simply replicated as closely as possible. This difference implies that relative factor intensities across sectors do not vary much in this chapter.[33] Other differences in the two models relate to rigidities; the version in this chapter has less rigidities on both labour and goods markets (see also Section 4). Third, the experiments differ. Chapter 8 asks how inequality in the OECD would look when the non-OECD would have been isolated from the OECD, whereas the issue here is what further integration would imply for inequality.

5.3 Sensitivity analysis

In this section we examine how sensitive the results are for the assumption that the R&D intensities in the non-OECD remain as low as they currently are. We do so by implementing a relation between the R&D intensity and GDP per capita.[34]

Increasing R&D intensities in the non-OECD

So far we have assumed that the ratio of sectoral R&D expenditures to value added is constant in the simulation period. This seems to be reasonable for the developed regions in which R&D expenditures do not vary substantially over time, but not for the developing regions. The analysis of the R&D data of UNESCO (1998) shows that R&D intensities increase as countries become more wealthy (see Appendix A). According to this analysis, the ratio of regional R&D expenditures to value added rises by 0.4 percentage-points if per capita income doubles. We analyse the GDP effects of the spillovers once we introduce this relation in our simulations. In the benchmark simulations presented above some spillovers were negative because the R&D content of the intermediate inputs decreased in the presence of trade liberalization. At the moment that the R&D intensities in the non-OECD regions increase, the R&D content of intermediate goods will raise. Table 9.11 shows the results in deviation from those in Table 9.7.

Column (4) shows that only the non-OECD regions are seriously affected compared to the case of constant R&D intensities in the non-OECD. The effects on the OECD regions are negligible. In the non-OECD regions, the sectoral spillovers are much larger. The sectoral spillovers are positive for all regions. For all regions, except China, the total sectoral effect – the sum of the columns (2) in Table 9.7 and 9.10 – is positive. And for China this negative effect is much smaller now. The effects of the international spillovers are ambiguous. However, the changes are fairly small.

Table 9.11 Deviations in the cumulative GDP effects of trade liberalization due to increasing R&D intensities in the non-OECD

Region	Own R&D (1)	Sectoral R&D (2)	International R&D (3)	Total (4)
United States	0.0	0.0	0.0	0.0
Western Europe	-0.1	0.0	0.0	0.0
Japan	0.0	0.1	0.0	0.0
Pacific OECD	-0.1	0.0	0.2	0.1
Eastern Europe	0.0	0.3	0.1	0.3
Former Soviet Union	0.0	0.1	0.0	0.1
Latin America	0.0	0.5	0.2	0.7
Middle East & N. Africa	0.0	0.4	-0.2	0.2
Sub-Saharan Africa	0.0	0.4	0.2	0.6
China	0.3	0.5	0.2	0.9
South-East Asia	0.0	1.2	-0.1	1.1
South Asia & Rest	0.5	0.6	0.1	1.2

Note: All results are presented in deviation from Table 9.7.
Source: WorldScan simulations.

6. CONCLUSIONS

Do R&D and R&D spillovers provide a link between openness and growth? The answer to that question is affirmative according to our analysis. The introduction of R&D in our AGE model always increases the effect of trade liberalization. The size of the effect depends heavily on specialization patterns and changes in that pattern due to trade liberalization. A more intense relation with one sector or region often implies a less intense relation with other sectors or regions. A change in the intermediate input structure or trade pattern only raises productivity if it is a change towards R&D-intensive sectors or regions.This is one of the main conclusions of this chapter.

Although R&D enlarges the benefits from trade liberalization, the effects are region- and sector specific. Here the value added of our AGE model, WorldScan, comes in. It allows for regional and sectoral detail. Therefore, we can model inter- and intra-regional spillovers from R&D. Sectors and regions face a trade-off with respect to these spillovers. R&D spillovers can be obtained by producing R&D-intensive and spillover-intensive goods domestically or by importing them. In the former case the intra-regional (sectoral) spillovers are important. Regions which have a comparative advantage in R&D-intensive sectors rely on this mechanism. As producing R&D-intensive goods turns out skill- and capital-intensive, the intra-regional spillovers are important for Western Europe, Eastern Europe and Japan.
Other regions which specialize in agriculture or that are not skill- and capital-rich obtain the R&D spillovers by the international linkages. For some regions the gains from trade liberalization are even reduced by negative sectoral spillovers. In the process of trade liberalization, these regions specialize in R&D-extensive products. As a consequence the R&D-intensive sectors move away to other regions.

These results do not advocate policies that hamper trade, as the gains from trade liberalization (with R&D in the model) are still (more) positive. Two policy options, however, find support in the analysis. The first policy option is 'infant-industry' protection. Selective trade liberalization – protecting the industries that are important for intra-sectoral spillovers – might help to benefit from the sizeable intra-national spillovers along the international spillovers from industries that are less R&D intensive. Such a policy, however, conflicts with WTO arrangements and is therefore further ignored. The second policy option is stimulating R&D. If regions increase R&D expenditures, the negative spillovers of trade liberalization reduce or even disappear. Our analyses showed that the sectoral spillovers then become more important. A policy which stimulates R&D does not necessarily have to be directed to the R&D-intensive sectors. It makes sense to stimulate those sectors which produce goods that are often used as intermediate input. However, as those goods are often imported, it could make more sense to target R&D-stimulating policy to those sectors which are also often used domestically as intermediate goods, such as services.

NOTES

1. That such a link exists is partly based on casual observations about the effect of an isolationalist policy on technological sophistication and partly on empirical work by, for example, Sachs and Warner (1995). The empirical relation is, however, controversial. See Rodríguez and Rodrik (1999).
2. Grossman and Helpman (1991) discuss the effects of openness on research productivity and the reduction of duplicationary research. See Rivera-Batiz and Romer (1991) for the scale effect in research. The market-size effect is discussed in Acemoglu (1998).
3. How important trade partners are is still open to debate. Lichtenberg and Pottelsberghe de la Potterie (1998) argue that FDI flows matter.
4. Numerical estimates in AGE modelling consistently show low welfare increases due to trade liberalization in case the models are of the static CRS type.
5. There exist, to our knowledge, only few other studies that perform similar exercises. Van Meijl and van Tongeren (1999) propose an absorption-capacity based spillover measure and test the numerical consequences of that by bringing the spillover measure to the GTAP data and model. Rutherford and Tarr (1998) develop an R&D-based CGE model for the small open economy. Their model, however, remains highly stylized and is simulated without a clear link to the data. We add to these contributions by estimating the relations present in the data and implementing them in a calibrated model that generates transition dynamics.
6. For this point it is crucial that knowledge spillovers are tied to imports. As such, nothing precludes spillovers related to exports (learning by competition on the international market); here we choose, however, to follow the lines set out in the theory and empirical work discussed above.
7. See Grossman and Helpman (1991) for a thorough theoretical analysis of these issues.
8. Note, for later reference, that the assumption of a production function that is homogeneous of degree one implies that TFP is homogenous of degree zero with respect to the conventional inputs Q and L and hence independent of the scale of the economy.
9. At first sight this might seem inconsistent; paying for the R^{DD} factor would lead to losses in a competitive market where L and Q jointly enter with constant returns. However, the expenditures for R^{DD} are included in Q and L already. In the estimation below the elasticity on R^{DD} is to be interpreted as an excess elasticity; hence the effect of R^{DD} should be interpreted as an intra-sectoral externality in order to maintain the constant-returns

assumption.

10. In Section 3.1 it will become clear that we derive TFP data in a slightly different way, using a translog function.
11. This point can be made for R&D stocks too. However, the data and conceptual problems to account appropriately for the growth in the R&D stock in a growth-accounting sense are huge. Moreover, appropriate instruments for R&D are hard to imagine. See also Barro (1999).
12. See Keller (1997) for an explicit derivation.
13. Ideally one wants to estimate a separate spillover elasticity for each knowledge source. For such an approach data availability is the limiting factor.
14. A micro foundation for this assumption is provided in Peretto and Smulders (2002).
15. In continuous time: $\hat{R}=RD/Y\bullet Y/R-\delta$. If δ is zero we get $RD/Y=R/Y\bullet\hat{R}$.
16. Assume a world with three countries, with domestic R&D capital stocks (R) for countries 2 and 3 equal to $R_2 = 10$ and $R_3 = 20$ respectively. Then, if country 1 imports 10 from country 2 and 10 from country 3, its foreign R&D capital stock (R^F) should be calculated as follows (assume the weights sum to unity): R^F=(10/20)10+(10/20)20=15. If we assume that countries 2 and 3 merge into one single country, the foreign R&D capital stock of country 1 becomes (with the same trade flows as before): R^F=(20/20)30=30, which is twice as large as the foreign R&D capital stock estimated from two distinct countries. That is, the foreign capital stock suffers from an aggregation bias. This example is taken directly from Lichtenberg and Pottelsberghe de la Potterie (1996). What is done in the remainder of the main text can be explained by this example as follows: given that the theoretically relevant construct suffers from an aggregation bias, the estimated coefficients for the spillover stock should be allowed to differ between countries of different sizes.
17. In the estimations we do not have time-series variation in the weights.
18. The data on wage-income shares, value added, capital and employment are directly available from the ISDB database, for details see OECD (1999b). The data series for hours worked are not sector specific. The data series for hours worked for Italy is extended by assuming no change after 1985 whereas for Denmark we use Maddison (1991, Table C9) data (with linear intrapolation).
19. For the US, agriculture and mining is included in services. We assume that the R&D intensity is equal in these three sectors in the US. For more details refer to Lejour and Nahuis (2000).
20. With respect to the sectoral aggregation, we assume that the services and trade and transport sectors have the same R&D intensity, which is approximated by the R&D intensity of services in the OECD data. For the manufacturing sectors we aggregate the sectors S3100, S3200, S3300 and S3900 to the consumer goods sector. The capital goods sector is simply S3800, while the energy-intensive goods sector consists of the other disaggregated manufacturing sectors, S3400 to S3700.

21. For our purposes we face two problems. First, the data include all expenditures on R&D, not only business enterprise. Second, the data do not include a sectoral division. The first problem is solved by using Table 5.6 from UNESCO (1998). This statistical yearbook provides information on the R&D expenditures by sector of performance. We interpret the productive sector in this table as business enterprises. The second problem is solved by applying the average OECD relative R&D intensities to the non-OECD regions. These relative intensities are multiplied by the business enterprise R&D ratio in Table 9.3. The results are shown in Table A.2.
22. For that aim, it is problematic that we reduce the variation in the data considerably by aggregating the data to our desired sectoral and regional level. The relations we establish are confirmed by work of Coe and Helpman (1995), Park (1995), Keller (1997) and Verspagen (1997). Later in this section we will briefly discuss this research.
23. This is a very crude approximation, based on the fact that the sum of the input coefficients is 0.1 to 0.4 (recall that we exclude intra-sectoral use). As, in many estimations present in the literature, the coefficients are adjusted such that they sum to one, we blow up our coefficient by a factor of 2.5 to 10. On average the adjustment factor is approximately 5.
24. An Armington trade specification amends the neo-classical trade theory. This is to explain two-way trade and to allow for market power to determine trade patterns in the medium run, while allowing for Heckscher-Ohlin mechanisms in the long run.
25. CPB (1999) provides more details of the globalization scenario. This scenario is akin to the high growth scenario which CPB and OECD have constructed for their collaborative study on globalization and the consequences for the OECD countries (OECD, 1997).
26. In the estimations, we assumed that R&D stocks did not depreciate. Some sensitivity analysis of our simulations with respect to the assumed depreciation rate is presented in Appendix D. In Appendix D we set δ equal to zero. The qualitative results are not altered.
27. In the policy simulations we take the calculated increases in A as exogenous.
28. Appendix D contains two additional sensitivity analyses: (i) the analysis in the main text redone with an R&D-capital depreciation rate of zero and (ii) the analysis is redone with modified elasticities of TFP to R&D spillovers.
29. The substantial GDP effects can partly be explained by the fact that we assume that the consumer preferences for a certain variety (in the Armington demand functions) depend positively on the share in global production of the region in which the variety is produced. In the trade-liberalization simulation this works as a multiplier: an increase in a region's GDP (share in the world) induces higher demand and thereby easier market penetration, which again leads to an increase in GDP and so on.
30. Table 9.7 presents the GDP effects of trade liberalization. These effects also include terms-of-trade effects. Alternatively, we could present the effects on the volume of consumption. The effects in the initial policy simulation without R&D in the model are different. The consumption gains for the Asian regions are substantially lower than the percentage gains in GDP in Table 9.7. The effects of introducing R&D in these simulations is the same as above, however. The same regions have relatively large sectoral or international spillovers.

All conclusions thus hold whether the analysis is based on GDP effects or consumption effects.

31. The numbers in Table 9.8 are summary statistics. The first two columns in Table 9.8 give an indication for the magnitudes in the third column. The latter gives an indication for the size of the sectoral spillovers. However, there is no one-to-one relation with column (3) in Table 9.7. For example, the sectoral spillovers for the United States are positive, while the relative change in the total R&D stock is negative. This can be explained by different R&D-stock elasticities between sectors.
32. As the numbers in Table 9.8, the numbers in Table 9.9 are only indicators of the size of the international spillovers. A one-to-one mapping between the indicators and Table 9.9 is not possible because of different R&D-stock elasticities for sectors and regions.
33. Nahuis (1999) extensively discusses why aggregation of data with factor-proportion theory in mind is preferable for the type of exercises discussed here. See also Leamer (1994a).
34. In Appendix D, we check the robustness of our results for the size of estimated coefficients and present the results for an R&D-capital depreciation rate of zero.

APPENDICES

A. R&D data

Table A.1 R&D intensities for various countries at a sectoral level in 1990 (ISIC2 classification)

Sectoral R&D	Agriculture	Raw materials	Food, tobacco	Textiles	Wood	Paper
Canada	0.18	0.46	0.48	0.73	0.65	0.76
Denmark	0.50	0.00	1.40	0.43	0.26	0.23
Finland	0.42	0.42	2.71	1.17	0.90	2.05
France	0.42	0.88	0.94	0.40	0.17	0.27
Germany	0.93	0.93	0.43	0.64	0.67	0.34
Italy	0.00	0.00	0.26	0.04	0.04	0.01
Japan	0.10	2.65	1.91	3.52	0.00	2.57
Norway	0.97	0.97	1.46	0.89	0.75	0.68
Sweden	0.54	0.54	1.61	1.21	0.18	2.14
UK	0.75	1.01	1.32	0.27	0.33	0.34
US	0.53	0.53	1.27	0.57	0.52	0.88
Sectoral R&D	**Chemicals, rubbers, plas.**	**Stone and clay**	**Basic metals**	**Fabricated metals**	**Other man.**	**Services**
Canada	4.59	0.51	3.01	7.07	0.96	0.41
Denmark	8.92	1.95	4.69	5.65	13.36	0.50
Finland	9.10	2.06	3.79	7.75	0.68	0.32
France	7.61	1.59	2.52	11.08	0.89	0.17
Germany	8.12	1.63	1.02	9.34	1.29	0.12
Italy	5.80	0.24	1.58	5.89	0.27	0.13
Japan	12.88	4.92	4.70	10.64	0.36	0.12
Norway	7.44	0.00	6.87	8.78	0.28	0.34
Sweden	13.44	1.65	3.74	14.04	3.13	0.44
UK	12.09	1.28	0.63	11.06	0.69	0.48
US	9.37	2.46	1.72	15.22	2.99	0.53

Source: OECD (1999a, 1999b) and own calculations.

Sectoral R&D intensities in WorldScan for the non-OECD

Table A.2 Sectoral R&D intensities in WorldScan for the non-OECD as ratio of sectoral value added

Sectoral R&D intensities	Agri-culture	Raw materials	Consumer goods	Energy goods	Capital goods	Services	Aver-age
Eastern Europe	0.11	0.16	0.22	1.26	2.79	0.07	0.42
Former Soviet Union	0.12	0.19	0.25	1.45	3.22	0.08	0.49
Latin America	0.04	0.06	0.08	0.48	1.07	0.03	0.16
Middle East	0.08	0.12	0.16	0.94	2.10	0.05	0.32
Sub-Saharan Africa	0.08	0.12	0.17	0.96	2.13	0.05	0.32
China	0.05	0.07	0.10	0.58	1.29	0.03	0.19
South-East Asia	0.23	0.35	0.48	2.77	6.13	0.16	0.93
South Asia & Rest	0.05	0.07	0.09	0.55	1.21	0.03	0.18

Source: UNESCO (1998, 1999), OECD (1999a, 1999b) and own calculations.

Trend in R&D expenditures

In WorldScan, we model the R&D stock as a capital stock: the R&D stock in the previous period (corrected for depreciation) plus the outlays in the current period. These outlays are derived from an exogenous ratio of value added to R&D investment. Our analysis of the UNESCO (1998) data shows that low-income countries spend relatively less on R&D than high-income countries. Figure A.1 shows the correlation between the R&D expenditures /GDP ratio and log GNP per capita (Worldbank, 1997) for about 85 countries in 1995.

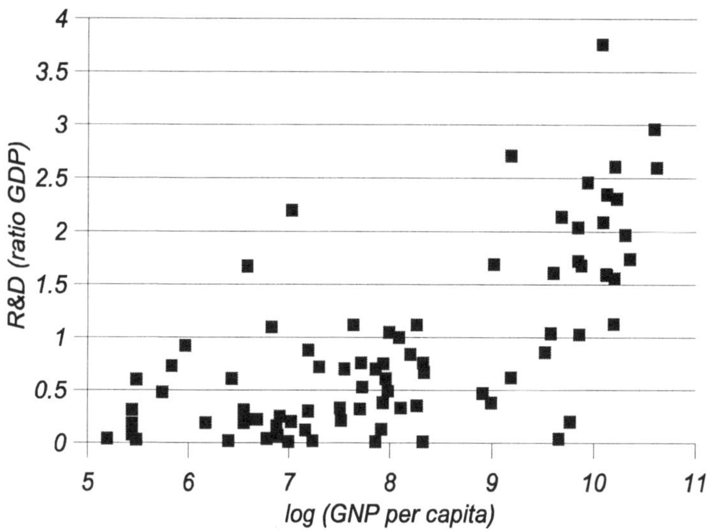

Figure A.1 R&D intensity and income per capita

From this figure we see a positive relationship between economic development (measured by GNP per capita) and expenditures on R&D. This relationship follows also from a simple OLS regression:

$$RD/GDP = -2.13 + 0.38 \log (\text{GNP per capita})$$
$$(0.41) \quad (0.05)$$

This regression shows that if GNP per capita doubles, the R&D-GDP ratio increases by 0.38 percentage-points.[1] So, if low-income countries develop, their R&D-GDP ratio will increase substantially. We use this result in the sensitivity analysis reported on in the main text.

B. Sensitivity analysis for the regression results

This appendix contains two tables with regression results to corroborate the robustness of the findings in the main text. First, we present analogous regressions to those in the main text, but here for manufacturing only. The second table includes some regressions in 'levels'.

We re-evaluate the results for the manufacturing sectors only as we are mainly interested in international spillovers and the bulk of the international trade concerns the manufacturing sectors.[2]

Table B.1 OLS results for manufacturing; equation (9.10). Dependent variable is TFP growth.[†]

Variable	(I) Direct effect	(II) Direct + indirect effect	(III) Domestic and Total	(IV) Domestic and Foreign
DD	0.158**	0.124*	0.096	0.162**
	[0.069]	[0.07]	[0.073]	[0.079]
ID		2.408**	2.943***	3.797***
		[1.045]	[1.11]	[1.175]
TF			0.541	
			[.386]	
DF[a]				0.107
				[0.436]
IF[b]				10.627**
				[4.844]
R^2 (adjusted)	0.02	0.04	0.05	0.06
N	216	216	216	216

Notes: [†] Sample period is 1974–91, 3 sectors and 4 regions. All regressions include a constant. The explanatory variables are lagged by one year. Standard errors are given in parentheses under the estimates.

*, **, and *** denote statistical significance at the 10 per cent level, the 5 per cent level, and the 1 per cent level, respectively.

[a] Direct Foreign: only the diagonal elements of import IO matrix are non-zero.

[b] Indirect Foreign: the diagonal elements of import IO matrix are zero.

The results are not dramatically different from those in the main text. Surprisingly, the regression in column (IV) in Table B.1 indicates that for manufacturing sectors only the indirect foreign spillovers are important.

Table B.2 OLS results for aggregate model; equation (B.1). Dependent variable is ln(TFP).†

Variable	(I) Direct effect	(II) Direct + indirect effect	(III) Domestic and Foreign
DD	0.189***	0.184***	0.159***
	[0.026]	[0.025]	[0.027]
ID		1.14***	1.538***
		[0.378]	[0.411]
TF			0.410**
			[0.173]
R^2 (adjusted)	0.78	0.78	0.78
N	456	456	456

Notes: † Sample period is 1973–91, 6 sectors and 4 regions. All regressions include sector-specific constants and a time trend. Standard errors are given in parentheses under the estimates.
*, **, and *** denote statistical significance at the 10 per cent level, the 5 per cent level, and the 1 per cent level, respectively.

In Table B.2 we present analogous regressions to those in the main text, but now estimated in 'levels'. Columns (I) to (III) in Table B.2 are analogous to columns (I) to (III) in Table 9.4 in the main text. The estimations are based on the integrated version of equation (9.10):

$$\begin{aligned}\log F_{ikt} = \; & ct + \beta^{DD} \sum_{\tau=0}^{t} \frac{RD_{ik\tau}}{Y_{ik\tau}} + \beta^{ID} \sum_{j \neq i}^{I} \omega_{ijk} \sum_{\tau=0}^{t} \frac{RD_{jk\tau}}{Y_{jk\tau}} \\ & + \beta^{F} \sum_{l \neq k}^{K} \sum_{i}^{I} n_{ijkl} \sum_{\tau=0}^{t} \frac{RD_{jk\tau}}{Y_{jk\tau}} + u_{ikt}\end{aligned} \qquad (B.1)$$

This is an expression in 'levels'. Given our aim, we do not extensively discuss econometric issues related to the estimation of equations in levels with variables that are non-stationary.[3] The elasticities estimated here are somewhat lower than those in the main text. The coefficient on foreign R&D now differs significantly from zero.

The results in the main text are not very sensitive for the lag structure – of one year– that we impose. We do not report these regressions. These are available upon request.

C. Calibration

From estimation to implementation in WorldScan

This appendix explains how the estimated equation is translated to the necessary model input. Helpful for this appendix is to recall that we estimated an equation in growth rates and implement an equation in levels in the CGE model and that the estimates are for an equation with R&D intensities whereas R&D stocks are introduced in the CGE model.

If we differentiate the equation in levels (see equation (9.3)) with respect to time we get

$$\hat{F}_{ik} = \hat{A}_{ik} + \gamma_{ik}^{DD}\hat{R}_{ik}^{DD} + \gamma_{ik}^{ID}\hat{R}_{ik}^{ID} + \gamma_{ik}^{F}\hat{R}_{ik}^{F} \; . \tag{C.1}$$

As in the base-line simulation all variables - except of course the γ parameters - are time varying we drop time indices where it leads to no confusion.[4] A hat denotes a growth rate. The variables in equation (C.1) are defined as follows:

$$R_{ik}^{ID} = \sum_{j \neq i}^{I} \omega_{jik} R_{jk} \; , \tag{C.2}$$

and

$$R_{ik}^{F} = \sum_{l \neq k}^{K} \sum_{j}^{I} n_{jilk} R_{jl} \ . \tag{C.3}$$

where ω denotes the IO-coefficient from sector j to i, and n the sectoral bilateral trade coefficient from country l to k. Knowledge stocks are weighted sums of other sectors' and countries' R&D stocks. Differentiation of equation (C.2) with respect to time leads to (one can follow an analogous procedure for the own R&D stock and the foreign R&D stock):

$$\hat{R}_{ik}^{ID} = \sum_{j \neq i}^{I} \frac{\omega_{jik} \Delta R_{jk}}{R_{ik}^{ID}} \ . \tag{C.4}$$

The coefficient γ, times the growth rate of the R&D stock of indirect spillovers (see equation (C.1)) has to be equal to the one in the estimated equation (equation (9.4) in the main text). So,

$$\gamma_{ik}^{ID} \hat{R}_{ik}^{ID} = \beta^{ID} \sum_{j \neq i}^{I} \omega_{jikt} \, rd_{jkt} \ , \quad \forall \ i,k \ . \tag{C.5}$$

Use that:[5]

$$\frac{\Delta R_{ik}}{R_{ik}} \equiv \frac{RD_{ik}}{R_{ik}} \equiv \frac{rd_{ik} Y_{ik}}{R_{ik}} \ , \tag{C.6}$$

and (C.4) to write equation (C.5) as:

$$\gamma_{ik}^{ID} \sum_{j \neq i}^{I} \frac{\omega_{jik} \Delta R_{jk}}{R_{ik}^{ID}} = \beta^{ID} \sum_{j \neq i}^{I} \frac{\omega_{jik} \Delta R_{jk}}{Y_{jk}} \quad \forall \ i,k \ . \tag{C.7}$$

Integrate both sides of equation (C.7) to get:

$$\gamma_{ik}^{ID} \sum_{j \neq i}^{I} \frac{\omega_{jik} R_{jk}}{R_{ik}^{ID}} = \beta^{ID} \sum_{j \neq i}^{I} \frac{\omega_{jik} R_{jk}}{Y_{jk}} \quad \forall\, i,k \,. \tag{C.8}$$

Use the definition of R_{ik}^{ID} (equation (C.2)) to see that the left-hand side of (C.8) is equal to γ_{ik}^{ID}. Thus the parameters of interest are the sector-country-specific corrected estimates:

$$\gamma_{ik}^{ID} = \beta^{ID} \sum_{j \neq i}^{I} \frac{\omega_{jik} R_{jk}}{Y_{jk}} \quad \forall\, i,k \,. \tag{C.9}$$

An analogous procedure is to be followed to obtain the other coefficients for equation (C.1).

D. Simulations without depreciation and with variation in the estimated coefficients

This appendix discusses two additional sensitivity analyses of the results presented in Section 5.2 of the main text.

No depreciation

In the main text we apply a depreciation rate of R&D capital of 5 per cent. In the estimations we assumed, however, that the depreciation rate was 0 per cent. Table D.1 reports the differences between such an experiment and the results in the main text.

Table D.1 Cumulative GDP effects of trade liberalization in 2020 without depreciation

Region	Own R&D	Sectoral R&D spillovers	International R&D spillovers	Relative GDP increase due to R&D
	(2)	(3)	(4)	(5)
United States	0.1	0.5	0.7	83.6
Western Europe	0.6	3.1	0.2	227.5
Japan	1.4	10.5	0.3	524.2
Pacific OECD	0.3	0.6	- 0.4	13.0
Eastern Europe	0.3	1.3	1.2	55.2
Former Soviet Union	0.2	0.3	0.4	53.8
Latin America	0.4	0.9	- 0.1	12.7
Middle East & N. Africa	0.4	0.6	5.0	125.1
Sub-Saharan Africa	0.1	-0.4	0.8	11.4
China	0.2	- 0.7	1.2	4.4
South-East Asia	1.8	7.8	2.7	82.5

Source: WorldScan simulations.

If one compares the results in Table D.1 with Table 9.7 in the main text, it follows that the GDP effects of trade liberalization are in general larger. This is in particularly the case for those regions in which the extra effects of incorporating R&D are fairly large in the benchmark simulation; the ones with large sectoral R&D spillovers.

Sensitivity to the estimated coefficients

The importance of the sectoral and international spillovers depends on the size of the R&D elasticities in the TFP function (see equation (9.12)). We change these elasticities to get some idea of the sensitivity of changes in these elasticities on the GDP effects of trade liberalization. In the benchmark simulations our starting point is the estimated elasticities of own R&D, sectoral R&D and international R&D which are equal to 0.167, 2.636, and 0.618 respectively (see Table 9.4). These elasticities are modified such that the appropriate γs for every sector and region are implemented in the model. In this sensitivity analysis we assume that the estimated elasticities for the sectoral and international spillovers are equal. We use a value of 1.674, the average of the estimated elasticities. Table D.2 presents the deviations of the GDP effects of the simulations where the R&D elasticities are modified, compared to the simulations in Table 9.7. As only the GDP effects which are related to the sectoral and international spillovers are modified, Table D.2 presents only these deviations and the deviations in the total GDP effect.

Table D.2 Deviations of cumulative GDP effects of trade liberalization in 2020 (R&D elasticities are modified)

Region	Sectoral R&D spillovers (2)	International R&D spillovers (3)	Total
United States	-0.1	0.5	0.4
Western Europe	-1.0	0.2	-0.7
Japan	-3.5	0.3	-3.2
Pacific OECD	-0.2	0.4	0.2
Eastern Europe	-0.4	1.1	0.7
Former Soviet Union	-0.1	0.4	0.3
Latin America	-0.3	0.6	0.3
Middle East & N. Africa	-0.2	3.3	3.1
Sub-Saharan Africa	0.1	1.1	1.2
China	0.2	1.4	1.5
South-East Asia	-2.5	1.8	-0.7
South Asia & Rest	-0.2	0.9	0.7

Source: WorldScan simulations. All results are presented in deviation from Table 9.7.

Not surprisingly, Table D.1 shows that the effects of the sectoral spillovers are smaller, while those of the international spillovers are larger. Lejour and Nahuis (2000) more elaborately discuss the sensitivity analyses.

NOTES

1. Standard errors are reported between parentheses. The adjusted R^2 is 0.45.
2. Excluding the agriculture and raw materials sector is motivated by the fact that the R&D data for these sectors are far less reliable.
3. Estimating variables that are individually non-stationary is possible if the combination of variables is cointegrating. That is, if a linear combination of the variables exists which, in a regression, yields a stationary error term. The t-statistics should however be interpreted with caution. Moreover, the robustness of the Coe and Helpman results to the estimation method

is established by Engelbrecht (1997). Estimations in log difference yields similar and significant results to the estimation of the cointegrated relations. Kao, Chiang and Chen (1999) use modern panel-cointegration techniques to redo the Coe and Helpman estimations and confirm most findings; most t-statistics turn out considerably lower (some are reduced by half).

4. As motivated in the main text, we keep the IO-coefficients at the base-line level in the policy simulations.
5. We assume in equation (C.6) that R&D stocks do not depreciate. Here we follow the Terleckyj approach (1974).

References

Acemoglu, D. (1998), 'Why Do New Technologies Complement Skills? Directed Technical Change and Wage Inequality', *Quarterly Journal of Economics*, 113, 1055–89.

Adams, J.D. (2000), 'Endogenous R&D Spillovers and Industrial Research Productivity', NBER Working Paper No. 7484.

Aghion, P. and P. Howitt (1996), 'On the Macroeconomic Effects of Major Technological Change', Discussion paper.

Aghion, P. and P. Howitt (1998a), *Endogenous Growth Theory*, The MIT Press, Cambridge, Massachusetts, London.

Aghion, P. and P. Howitt (1998b), 'Market Structure and the Growth Process', *Review of Economic Dynamics*, 1, 276–305.

Allen, D.S. (1997), 'Where's the Productivity Growth (from the Information Technology Revolution)?', *Federal Reserve Bank of St. Louis Review*, 15–25.

Andolfatto, D. and G.M. MacDonald (1998), 'Technology Diffusion and Aggregate Dynamics', *Review of Economic Dynamics*, 1, 338–70.

Antonelli, C. (1995), *The Economics of Localized Technological Change and Industrial Dynamics*, Kluwer Academic Publishers, Dordrecht, The Netherlands.

Atkinson, A.B. and J.E. Stiglitz (1969), 'A New View of Technological Change', *Economic Journal*, 79, 573–8.

Autor, D.H., L.F. Katz and A.B. Krueger (1998), 'Computing Inequality: Have Computers Changed the Labour Market?', *Quarterly Journal of Economics*, 113, 1169–213.

Baily, M.N. (1993), 'Competition, Regulation, and Efficiency in Service Industries', *Brookings Papers on Economic Activity, Microeconomics*, 2,

71–130.

Baily, M.N. and H. Gersbach (1995), 'Efficiency in Manufacturing and the Need for Global Competition', *Brookings Papers on Economic Activity, Microeconomics*, 4, 307–347.

Baldwin, R.E. (1995), 'The Effect of Trade and Foreign Direct Investment on Employment and Relative Wages', NBER Working Paper No. 5037.

Baldwin, R.E. and G.G. Cain (1997), 'Shifts in U.S. Relative Wages: The Role of Trade, Technology and Factor Endowments', NBER Working Paper No. 5934.

Barro, R.J. (1999), 'Notes on Growth Accounting', *Journal of Economic Growth*, 4, 119–37.

Barro, R.J. and X. Sala-i-Martin (1995), *Economic Growth*, McGraw-Hill Inc., New York.

Bartel, A.P. and F.R. Lichtenberg (1987), 'The Comparative Advantage of Educated Workers in Implementing New Technology', *The Review of Economics and Statistics*, LXIX, 1–11.

Basu, S. and D.N. Weil (1996), 'Appropriate Technology and Growth', *Quarterly Journal of Economics*, 113, 1025–54.

Bayoumi T., D. T. Coe and E. Helpman (1999), 'R&D Spillovers and Global Growth', *Journal of International Economics,* 47, 399–428.

Bell, L.A. (1994), 'The Impact of Minimum Wages in Mexico and Columbia', Presentation at the World Bank Labor Markets Workshop (cited via Feenstra and Hanson, 1995).

Benassy, J.P. (1996), 'Taste for Variety and Optimum Production Patterns in Monopolistic Competition', *Economics Letters*, 52, 41–47.

Benassy, J.P. (1998), 'Is There Always Too Little Research in Endogenous Growth with Expanding Product Variety?', *European Economic Review*, 42, 61–9 .

Benhabib, J. and M.M. Spiegel (1994), 'The role of Human Capital in Economic Development. Evidence from Aggregate Cross–Country Data', *Journal of Monetary Economics*, 34, 143–73.

Berman, E., J. Bound and Z. Griliches (1994), 'Changes in the Demand for Skilled Labor within U.S. Manufacturing: Evidence from the Annual

Survey of Manufactures', *Quarterly Journal of Economics*, 109, 367–97.

Bhagwati, J. (1991), 'Free Traders and Free Immigrationists: Strangers or Friends?', Working Paper 20, Russell Sage Foundation, New York (cited via Lawrence and Slaughter, 1993).

Bhagwati, J. and V.H. Dehejia (1994), 'Freer Trade and Wages of the Unskilled – Is Marx Striking Again?', in J. Bhagwati, and M.H. Kosters (eds.) *Trade and Wages. Levelling Wages Down?,* The AIE Press.

Bishop, J. (1991), 'Achievement, Test Scores, and Relative Wages,', in M. Kosters (ed.), *Workers and Their Wages*, Washington, DC, AEI press.

Borjas, G. and R. Freeman (1992), *The Economic Effects of Immigration in Source and Receiving Countries*, Chicago University Press.

Borjas, G.J. and V.A. Ramey (1993), 'Foreign Competition, Market Power, and Wage Inequality: Theory and Evidence', NBER Working Paper No. 4556.

Bound, J. and G. Johnson (1992), 'Changes in the Structure of Wages in the 1980's: An Evaluation of Alternative Explanations', *American Economic Review*, 82, 371–92.

Bresnahan, T.F. and M. Trajtenberg (1995), 'General Purpose Technologies "Engines of Growth"?', *Journal of Econometrics*, 65, 83–108.

Broer, D.P. and B.J. Heijdra (1996), 'The Intergenerational Distribution Effects of the Investment Tax Credit under Monopolistic Competition', OCfEB Research Memorandum 9603, Rotterdam.

Brynjolfsson, E. (1996), 'The Contribution of Information Technology to Consumer Welfare', *Information Systems Research: A journal of the Institute of Management Sciences*, 7, 281–300.

Brynjolfsson, E. and L.M. Hitt (1997), 'Paradox lost? Firm-level Evidence on the Returns to Information Systems Spending', *Operations Research/ Management Science*, 37, 83–6.

Brynjolfsson, E. and L.M. Hitt (1998), 'Beyond the Productivity Paradox', *Communications of the Association for Computing Machinery*, 41, 49–55.

Caballero, R.J. and M.L. Hammour (1998), 'The Macroeconomics of Specificity', *Journal of Political Economy*, 106, 724–67.

Canton, E.J.F. and A. Meijdam (1997), 'Altruism and the Macroeconomic Effects of Demographic Changes', *Journal of Population Economics*, 10, 317–34.

Canton, E.J.F., H.L.F. de Groot and R. Nahuis (1999), 'Vested Interests and Resistance to Technology Adoption', CentER Discussion Paper No. 99106.

Caselli, F. (1999), 'Technological Revolutions', *American Economic Review*, 89, 78–102.

Chari, V.V. and H. Hopenhayn (1991), 'Vintage Human Capital, Growth, and the Diffusion of New Technology', *Journal of Political Economy*, 99, 1142–65.

Cheng, L.K. and E. Dinopoulos (1996), 'A Multisectoral General Equilibrium Model of Schumpeterian Growth', *Journal of Economic Dynamics and Control*, 20, 905–23.

Coe, D.T. and E. Helpman (1995), 'International R&D Spillovers', *European Economic Review*, 39, 859–87.

Cohen, W.M. and D.A. Levinthal (1989), 'Innovation and Learning: the Two Faces of R&D', *Economic Journal*, 99, 569–97.

Cohen, W.M., R.R. Nelson and J.P. Walsh (2000), 'Protecting Their Intellectual Assets: Appropriability Conditions and Why U.S. Manufacturing Firms Patent (or Not)', NBER Working Paper No. 7552.

Cooper, R.N. (1994), 'Foreign Trade, Wages and Unemployment', Harvard Institute of Economic Research DP No. 1701.

CPB (1999), *WorldScan: the Core version*, The Hague.

David, P.A (1989), 'Computer and Dynamo. The Modern Productivity Paradox in a Not-Too Distant Mirror', mimeo.

David, P.A. (1990), 'The Dynamo and the Computer: An Historical Perspective on the Modern Productivity Paradox', *AEA Papers and Proceedings,* 80, 355–61.

David, P.A. (1991), 'Computer and Dynamo. The Modern Productivity Paradox in a Not-Too Distant Mirror', in *Technology and Productivity: The Challenge for Economic Policy*, 1991, OECD, Paris.

Davis, S.J. (1992), 'Cross-Country Patterns of Change in Relative Wages', in: O.J. Blanchard and S. Fischer (eds), *NBER Macroeconomic Annuals 1992*, 239–300.

Davis, S.J. and R.H. Topel (1993), 'International Trade and American Wages in the 1980s: Giant Sucking Sound or Small Hiccup? Comment', *Brookings Papers on Economic Activity*, Microeconomics, 214–21.

de Groot, H.L.F. (1998), *Economic Growth, Sectoral Structure and Unemployment*, PhD thesis, CentER dissertation, Tilburg, The Netherlands.

de Groot, H.L.F. and R. Nahuis (1998), 'Taste for Diversity and Optimality of Economic Growth', *Economic Letters*, 58, 291–295.

Deardorff, A.V. and D.S. Hakura (1994), 'Trade and Wages – What are the Questions?', in J. Bhagwati and M.H. Kosters (eds), *Trade and Wages. Levelling Wages Down?*, The AIE Press.

Deardorff, A.V. and R.W. Staiger (1988), 'An Interpretation of the Factor Content of Trade', *Journal of International Economics*, 24, 93–107.

Diamond, P. (ed.) (1990), 'Growth/Productivity/Unemployment, Essays to Celebrate Bob Solow's Birthday', The MIT Press, Cambridge Massachusetts.

Diewert, W.E. (1976), 'Exact and Superlative Index Numbers', *Journal of Econometrics*, 4, 115–45.

DiNardo, J.E. and J.S. Pischke (1997), 'The Returns to Computer Use Revisited: Have Pencils Changed the Wage Structure Too?', *Quarterly Journal of Economics*, 112, 291–303.

Dinopoulos, E. and P. Thompson (1995), 'Is Endogenous Growth Empirically Relevant', University of Florida Working Paper No. 95–96–8.

Dinopoulos, E. and P. Thompson (1996), 'A Contribution to the Empirics of Endogenous Growth', *Eastern Economic Journal*, 22, 389–400.

Dixit, A.K. and R.S. Pindyck (1994), *Investment under Uncertainty*, Princeton University Press, Princeton.

Dixit, A.K. and J.E. Stiglitz (1977), 'Monopolistic Competition and Optimum Product Diversity', *American Economic Review*, 67, 297–308.

Dornbusch, R., S. Fischer and P.A. Samuelson (1977), 'Comparative Advantage, Trade, and Payments in a Ricardian Model with a Continuum of Goods', *American Economic Review*, 67, 823–39.

Dosi, G. (1988), 'Sources, Procedures and Microeconomic Effects of Innovation', *Journal of Economic Literature*, 26, 1120–71. Reprinted in C. Freeman (ed.), *The Economics of Innovation*, An Elgar Reference Collection, 1990, Edward Elgar, Cheltenham, UK.

Eaton, J. and S. Kortum (1996), 'Trade in Ideas: Patenting and Productivity in the OECD', *Journal of International Economics*, 40, 251–78.

The Economist (1998), 'BMW's British Bruises', 5 December.

Engelbrecht, H.J. (1997), 'International R&D Spillovers, Human Capital and Productivity in OECD Economies: An Empirical Investigation', *European Economic Review*, 41, 1479–88.

Ericksson, C. and T. Lindh (1997), 'Growth Cycles with Technology Shifts and Externalities', Uppsala University, Working Paper Series 1997:15.

Ethier, W.J. (1982), 'National and International Returns to Scale in the Modern Theory of International Trade', *American Economic Review*, 72, 389–405.

Ethier, W.J. (1984), 'Higher Dimensional Issues in Trade Theory', in R. Jones and P. Kenen (eds), *Handbook of International Economics*, 1, North-Holland.

Feenstra, R.C. and G.H. Hanson (1995), 'Foreign Investment, Outsourcing and Relative Wages', NBER Working Paper No. 5121.

Feenstra, R.C. and G.H. Hanson (1996), 'Globalisation, Outsourcing and Wage Inequality', NBER Working Paper No. 5424.

Felli, L. and F. Ortalo-Magné (1997), 'Technological Innovations: Slumps and Booms', CARESS Working Paper No. 97–17.

Finger, J.M. (1975), 'Trade Overlap and Intra-Industry Trade', *Economic Inquiry*, XIII, 581–9.

Freeman, R. (1995), 'Are Your Wages Set in Beijing?', *Journal of Economic Perspectives*, 9, 15–32.

Galor, O. and O. Moav (1998), 'Ability Biased Technological Transition, Wage Inequality and Growth', CEPR Discussion Paper No. 1972.

Gerschenkron, A. (1962), *Economic Development in Historical Perspective*, Harvard University Press, Cambridge, Massachusetts.

Geurts, B., A. Gielen, R. Nahuis, P. Tang and H. Timmer (1995), 'WorldScan, an Economic *World* Model for *Sc*enario *An*alysis', Paper for the Review of WorldScan, The Hague.

Goldin, C. and L.F. Katz (1996), 'Technology, Skill, and the Wage Structure: Insights from the Past', *American Economic Review*, 86, 252–7.

Gordon, R.J, (2000), 'Does the "New Economy" Measure Up to the Great Inventions of the Past?', *Journal of Economic Perspectives*, 14, 49–74.

Greenwood, J. and M. Yorukoglu, (1997), '1974', *Carnegie-Rochester Conference Series on Public Policy*, 46, 49–95.

Griliches, Z. (1994), 'Productivity, R&D, and the Data Constraint', *American Economic Review*, 84, 1–23.

Grossman, G.M. and E. Helpman (1991), *Innovation and Growth in the Global Economy,* The MIT Press, Cambridge, Massachusetts.

Gullickson, W. (1995), 'Measurement of Productivity Growth in U.S. Manufacturing', *Monthly Labour Review*, 118, 13–28.

Hanson, G.H. and A. Harrison (1995), 'Trade, Technology, and Wage Inequality', NBER Working Paper No. 5110.

Harrison, A. and A. Revenga (1995), 'The Effects of Trade Policy Reform: What Do We Really Know', NBER Working Paper 5225.

Helpman, E and P. Krugman (1985), *Market Structure and Foreign Trade*, The MIT Press.

Helpman, E. and A. Rangel (1999), 'Adjusting to New Technology: Experience and Training', *Journal of Economic Growth*, 4, 359–83.

Helpman, E. and M. Trajtenberg (1994), 'A Time to Sow and a Time to Reap: Growth Based on General Purpose Technologies', NBER Working Paper No. 4854.

Helpman, E. and M. Trajtenberg (1996), 'Diffusion of General Purpose Technologies', NBER Working Paper No. 5773.

Holmes, T.J. and J.A. Schmitz, Jr (1994), 'Resistance to Change and Trade between Areas', Research Department Staff Report 184, Federal Reserve Bank of Minneapolis.

Holtz-Eakin, D. and M.A. Lovely (1994), 'Technological Linkages, Market Structure, and Optimum Production Policies', NBER Working Paper No.

4779.

Howitt, P. (1998), 'Measurement, Obsolescence, and General Purpose Technologies', in E. Helpman (ed.), *General Purpose Technologies and Economic Growth*, The MIT Press, Cambridge Massachusetts.

Hsieh, C-T. (2002), 'What Explains the Industrial Revolution in East Asia? Evidence from Factor Markets', *American Economic Review*, 92, 502–26.

Jacobs, B., R. Nahuis and P.J.G. Tang (2002), 'Sectoral Productivity Growth and R&D Spillovers in the Netherlands', *De Economist*, 150,181–210.

Jaffe, A.B. (1999), 'The U.S. Patent System in Transition: Policy Innovation and the Innovation Process', NBER Working Paper No. 7280.

Jaffe, A.B., M. Trajtenberg and R. Henderson (1993), 'Geographic Localisation of Knowledge Spillovers as Evidenced in Patent Citations', *Quarterly Journal of Economics*, 103, 577–98.

Jones, C.I. (1995a), 'Time Series Tests of Endogenous Growth Models', *Quarterly Journal of Economics*, 110, 495–525.

Jones, C.I. (1995b), 'R&D-Based Models of Economic Growth', *Journal of Political Economy*, 103, 759–84.

Jones, C.I. (1996), 'Human Capital, Ideas, and Economic Growth', Stanford mimeo.

Jones, C.I., and J.C. Williams (2000), 'Too Much of a Good Thing? The Economics of Investment in R&D', *Journal of Economic Growth*, 5, 65–85.

Jorgenson, D.W. and K. Stiroh (1995), 'Computers and Growth', *Economics of Innovation and New Technology*, 3, 295–316.

Jovanovic, B. (1997), 'Learning and Growth', in D. Kreps and K. Wallis (eds), *Advances in Economics and Econometrics: Theory and Applications 2*, Seventh World Congress of the Econometric Society, Cambridge University Press, Cambridge.

Jovanovic, B. and Y. Nyarko (1996), 'Learning by Doing and the Choice of Technology', *Econometrica*, 64, 1299–310.

Jovanovic, B. and R. Rob (1990), 'Long Waves and Short Waves: Growth Through Intensive and Extensive Search', *Econometrica*, 58, 1391–409.

Kao, C., M-H. Chiang and B. Chen (1999), 'International R&D Spillovers:

An Application of Estimation and Inference in Panel Cointegration', *Oxford Bulletin of Economics and Statistics*, 61, No. 20, 691–710.

Katz, L.F. and K.M. Murphy (1992), 'Changes in Relative Wages, 1963–1987: Supply and Demand Factors', *Quarterly Journal of Economics*, 107, 35–78.

Keely, L.C. and D. Quah (1998), 'Technology in Growth', CEPR Discussion Paper No. 1901.

Keller, W. (1997), 'Trade and the Transmission of Technology', NBER Working Paper No. 6113.

Kiley, M.T. (1999), 'The Supply of Skilled Labour and Skill-Biased Technological Progress', *Economic Journal*, 109, 708.

Kim, J. and L.J. Lau (1994), 'The Sources of Economic Growth of the East Asian Newly Industrialized Countries', *Journal of the Japanese and International Economies*, 8, 235–71.

Komiya, R. (1967), 'Non–Traded Goods and the Pure Theory of International Trade', *International Economic Review*, 8, 132–52.

Kortum, S. (1993), 'Equilibrium R&D and the Patent-R&D Ratio: U.S. Evidence', *American Economic Review*, 83, 450–57.

Kortum, S. and J. Lerner (1998), 'Stronger Protection or Technological Revolution: What is behind the Recent Surge in Patenting?' *Carnegie Rochester Conference Series on Public Policy,* 48, 247–304.

Kosters, M.H. (1994), 'An overview of Changing Wage Patterns in the Labor Market', in J. Bhagwati and M.H. Kosters (eds), *Trade and Wages. Levelling Wages Down?*, The AIE Press.

Kremer, M. (1993), 'Population Growth and Technological Change: One Million B.C. to 1990', *Quarterly Journal of Economics*, 108, 681–716.

Krueger, A.B. (1993), 'How Computers Have Changed the Wage Structure: Evidence from Micro Data, 1984–1989', *Quarterly Journal of Economics*, 99, 99–118.

Krueger, A.B. (1997), 'Labor Market Shifts and the Price Puzzle Revisited', NBER Working Paper No. 5924.

Krueger, A.B. and L.H. Summers (1988), 'Efficiency Wages and the Inter–industry Wage Structure', *Econometrica*, 56, 259–93.

Krugman, P.R. (1994), *Peddling Prosperity. Economic Sense and Nonsense in the Age of Diminished Expectations*, W.W. Norton & Company, New York.

Krugman, P.R. (1995a), 'Growing World Trade: Causes and Consequences', *Brookings Papers on Economic Activity*, No 1, 327–62.

Krugman, P.R. (1995b), 'Technology, Trade, and Factor Prices', NBER Working Paper No. 5355.

Krugman, P.R. and R.Z. Lawrence (1993), 'Trade, Jobs and Wages', NBER Working Paper No. 4478.

Krusell, P. and J. Rios-Rull (1996), 'Vested Interests in a Positive Theory of Stagnation and Growth', *Review of Economic Studies*, 63, 301–29.

Lawrence, R.Z. (1994), 'Trade, Multinationals, & Labor', NBER Working Paper No. 4836.

Lawrence, R.Z. and M.J. Slaughter (1993), 'International Trade and American Wages in the 1980s: Giant Sucking Sound or Small Hiccup?', *Brookings Papers on Economic Activity* Microeconomics 2, 161–226.

Leamer, E.E. (1987), 'Paths of Development in the Three–Factor, *n*–Good General Equilibrium Model', *Journal of Political Economy*, 95, 961–99.

Leamer, E.E. (1994), 'Trade, Wages and Revolving Door Ideas', NBER Working Paper No. 4716.

Leamer, E.E. (1994a), 'Wage Effects of a U.S.–Mexican Free Trade Agreement', NBER Working Paper No. 3991.

Leamer, E.E. (1995), 'The Heckscher-Ohlin Model in Theory and Practice', Princeton Studies in International Finance No. 77.

Leamer, E.E. (1996a), 'In Search of Stolper–Samuelson Effects on U.S. Wages', NBER Working Paper No. 5427.

Leamer, E.E. (1996b), 'What is the Use of Factor Contents?', NBER Working Paper No. 5448.

Leamer, E.E. and P. Lundborg (1995), 'A Heckscher–Ohlin View of Sweden Competing in the Global Marketplace', NBER Working Paper No. 5114.

Lejour, A.M. and R. Nahuis (2000), 'Openness, Growth and R&D Spillovers: Uncovering the Missing Link?', CPB Research Memorandum No. 168.

Lejour, A.M. and P.J.G Tang (2000), 'Globalisation and Wage Inequality', CPB Research Memorandum No. 167.

Levy, F. and R.J. Murnane (1992), 'U.S. Earnings Levels and Earnings Inequality: A Review of Recent Trends and Proposed Explanations', *Journal of Economic Literature*, XXX, 1333–81.

Lichtenberg, F. and B. van Pottelsberghe de la Potterie (1996), 'International R&D Spillovers: A Re-examination', NBER Working Paper No. 5668.

Lichtenberg, F. and B. van Pottelsberghe de la Potterie (1998), 'International R&D Spillovers: A Comment', *European Economic Review*, 42, 1483–91.

Lipsey, R.G., C.Bekar and K. Carlaw (1998), 'What Requires Explanation?', in E. Helpman (ed.), *General Purpose Technologies and Economic Growth*, The MIT Press, Cambridge, Massachusetts.

Lloyd-Ellis, H. (1999), 'Endogenous Technological Change and Wage Inequality', *American Economic Review*, 89, 47–77.

Lucas, R.E., Jr (1988), 'On the Mechanics of Economic Development', *Journal of Monetary Economics*, 22, 3–42.

Machiavelli, N. (1961), *The Prince*, Penguin Books, Baltimore.

Machin, S. and J. van Reenen (1998), 'Technology and Changes in Skill Structure: Evidence From Seven OECD Countries', *Quarterly Journal of Economics*, 113, 1215–44.

Maddison, A. (1991), *Dynamic Forces in Capitalist Development. A Long-run Comparative View*, Oxford University Press, Oxford.

Mankiw, N.G. and M.D. Whinston (1986), 'Free Entry and Social Inefficiency', *Rand Journal of Economics*, Vol.17, 48–58.

Markusen, J.R. and A.J. Venables (1995), 'Multinational Firms and the New Trade Theory', NBER Working Paper No. 5036.

McDougall, R.A., A. Elbehri and T.P. Truong (1998), *Global Trade Assistance and Protection: The GTAP 4 Data Base,* Center for Global Trade Analysis, Purdue University, Purdue.

Mincer, J. (1991), 'Human Capital, Technology, and the Wage Structure: What do Time Series Show?', NBER Working Paper No. 3581.

Mokyr, J. (1990), *The Lever of Riches: Technological Creativity and*

Economic Progress, Oxford University Press, New York.

Mokyr, J. (1998), 'Knowledge, Technology, and Economic Growth During the Industrial Revolution', paper prepared for the conference on *Productivity and Standards of Living: Measurement, Modelling, and Market Behavior*, Groningen, September 23–25.

Morrison, C.J. and E.R. Berndt (1991), 'Assessing the Productivity of Information Technology Equipment in U.S. Manufacturing Industries', NBER Working Paper No. 3582.

Mulligan, C.B. and X. Sala-i-Martin (1993), 'Transitional Dynamics in Two-Sector Models of Endogenous Growth', *Quarterly Journal of Economics*, 108, 737–73.

Murphy, K.M. and F. Welch (1992), 'The Structure of Wages', *Quarterly Journal of Economics*, 107, 285–326.

Murphy, K.M., W.C. Ridell and P.M. Romer (1998), 'Wages, Skills, and Technology in the United States and Canada', in E. Helpman (ed.), *General Purpose Technologies and Economic Growth*, The MIT Press, Cambridge, Massachusetts.

Nadiri, M.I. (1993), 'Innovations and Technological Spillovers', NBER Working Paper No. 4423.

Nahuis, R. (1999), 'Global Integration and Wages in a General Equilibrium World Model: Contributions of WorldScan', in P. Brenton and J. Pelkmans (eds), *Global Trade and European Workers*, MacMillan Press.

Nahuis, R. and J. van de Ven (1999), 'Communications: Endogenous Growth in a Small Open Economy: Any Policy Lessons?', *De Economist*, 147, 353–60.

OECD (1985), *The Cost and Benefit of Protection*, Paris.

OECD (1993), *Employment Outlook 1993*, Paris

OECD (1997), *The World in 2020: Towards a New Global Age*, Paris.

OECD (1999a), *ANBERD 98 International Sectoral Data Base, User's Guide*, Paris.

OECD (1999b), *ISDB 98 International Sectoral Data Base, User's Guide*, Paris.

Oliner, S.D. and D.E. Sichel (1994), 'Computers and Output Growth

Revisited: How Big is the Puzzle?', *Brookings Papers on Economic Activity*, No. 2, 273–334.

Olson, M. (1982), *The Rise and Decline of Nations: Economic Growth, Stagflation, and Social Rigidities*, Yale University Press, New Haven.

Oosterbeek, H. (1997), 'Returns from Computer Use: A Simple Test on the Productivity Interpretation', *Economic Letters*, 55, 273–7.

Parente, S.L. (1995), 'A Model of Technology Adoption and Growth', *Economic Theory*, 6, 405–20.

Parente, S.L. and E.C. Prescott (1994), 'Barriers to Technology Adoption and Development', *Journal of Political Economy*, 102, 298–321.

Park, W.G. (1995), 'International R&D Spillovers and OECD Economic Growth', *Economic Inquiry*, 33, 571–91.

Peretto, P.F. (1996), 'Sunk Costs, Market Structure, and Growth', *International Economic Review*, 37, 895–923.

Peretto, P.F. (1998), 'Technological Change and Population Growth', *Journal of Economic Growth*, 3, 283–311.

Peretto, P.F. (1999), 'Cost Reduction, Entry, and the Interdependence of Market Structure and Economic Growth – A Reformulation', *Journal of Monetary Economics*, 43, 173–95.

Peretto, P.F. and J.A. Smulders (2002), 'Technological Distance, Growth and Scale Effects', *Economic Journal*, 112, 603–24.

Persson, T. and G. Tabellini (1994), 'Is Inequality Harmful for Growth?', *American Economic Review,* 84, 600–621.

Prescott, E.C. (1998), 'Needed: A Theory of Total Factor Productivity', *International Economic Review*, 39, 525–51.

Reich, R.B. (1992), *The Work of Nations, Preparing Ourselves for the 21st Century*, Vintage Books, New York.

Rivera-Batiz, L.A. and P.M. Romer (1991), 'Economic Integration and Endogenous Growth', *Quarterly Journal of Economics*, 106, 531–55.

Rodríguez, F. and D. Rodrik (1999), 'Trade Policy and Economic Growth: A Skeptics' Guide to the Cross-National Evidence', NBER Working Paper No. 7081.

Romer, P.M. (1986), 'Increasing Returns and Long–run Growth', *Journal of*

Political Economy, 94, 1002–37.

Romer, P.M. (1987), 'Crazy Explanations for the Productivity Slowdown', in S. Fischer (ed.), *NBER Macroeconomic Annual 1987*, 163–210.

Romer, P.M. (1990), 'Endogenous Technological Change', *Journal of Political Economy*, 98, S71–S103.

Romer, P.M. (1993), 'Idea Gaps and Object Gaps in Economic Development', *Journal of Monetary Economics*, 32, 543–73.

Rosen, S. (1981), 'The Economics of Superstars', *American Economic Review*, 71, 845–58.

Rustichini, A. and J.A. Schmitz (1991), 'Research and Imitation in Long-run Growth', *Journal of Monetary Economics*, 27, 271–92.

Rutherford, T.F. and D.G. Tarr (1998), 'Trade Liberalization and Endogenous Growth in a Small Open Economy: A Quantitative Assessment', *Fondazione Eni Enrico Mattei Note di Lavoro*, 60.

Sachs, J.D. and H.J. Shatz (1994), 'Trade and Jobs in U.S. Manufacturing', *Brookings Papers on Economic Activity*, 1–84.

Sachs, J.D. and A.M. Warner (1995), 'Economic Reform and the Process of Global Integration', *Brookings Papers on Economic Activity*, 1–95.

Saint-Paul, G. and T. Verdier (1996), 'Inequality, Redistribution and Growth: A Challenge to the Conventional Political Economy Approach', *European Economic Review*, 40, 719–28.

Sakurai, N., G. Papaconstantinou and E. Ioannidis (1997), 'Impact of R&D and Technology Diffusion on Productivity Growth: Empirical Evidence for 10 OECD Countries', *Economic Systems Research*, 9, 81–109.

Schumpeter, J.A. (1939), *Business Cycles: A Theoretical, Historical, and Statistical Analysis of the Capitalist Process*, McGraw-Hill, New York.

Scott, M.F. (1989), *A New View of Economic Growth*, Clarendon Press, Oxford.

Smulders, J.A. (1994), *Growth, Market Structure and the Environment*, Phd. Thesis, Tilburg University, Hilvarenbeek.

Smulders, J.A and Th. van de Klundert (1995), 'Imperfect Competition, Concentration and Growth with Firm-specific R&D', *European Economic Review*, 39, 139–60.

Solow, R.M. (1956), 'A Contribution to the Theory of Economic Growth', *Quarterly Journal of Economics*, 70, 65–94.

Solow, R.M. (1957), 'Technical Change and the Aggregate Production Function', *Review of Economics and Statistics*, 39, 312–20.

Solow, R.M. (1987), 'We'd Better Watch Out', *New York Times Book Review*, July 12.

Spence, M. (1976), 'Product Selection, Fixed Costs, and Monopolistic Competition', *Review of Economic Studies*, 43, 217–35.

Stein, J.C. (1997), 'Waves of Creative Destruction: Firm-Specific Learning-by-doing and the Dynamics of Innovation', *Review of Economic Studies*, 64, 265–88.

Stiglitz, J.E. (1987), 'Learning to Learn, Localized Learning and Technological Progress', in P. Dasgupta and J. Stiglitz (eds), *Economic Policy and Technological Performance*, Cambridge University Press, Cambridge.

Stolper, W. and P. Samuelson (1941), 'Protection and Real Wages', *Review of Economic Studies*, 9, 58–73.

Terlecky, N. (1974), *Effects of R&D on the Productivity Growth of Industries: An Exploratory Study*, National Planning Association, Washington DC.

Thompson, P. and D. Waldo (1994), 'Growth and Trustified Capitalism', *Journal of Monetary Economics*, 34, 445–62.

UNESCO (1998), *Statistical Yearbook Table 5.6*, Paris.

UNESCO (1999), *Statistical Yearbook Table III.1*, Paris.

van de Klundert, Th. (1999), 'Economic Development in Europe: 1960–1996', in M.M.G. Fase, W. Kanning and D.A. Walker (eds), *Economics, Welfare Policy and the History of Economic Thought, Essays in Honour of Arnold Heertje*, Edward Elgar, Cheltenham, UK.

van de Klundert, Th. and J.A. Smulders (1992), 'Reconstructing Growth Theory: A Survey', *De Economist*, 140, 177–203.

van de Klundert, Th. and J.A. Smulders (1997), 'Growth, Competition and Welfare', *Scandinavian Journal of Economics*, 99, 99–118.

van Meijl, H. and F. van Tongeren (1999), 'Endogenous International

Technology Spillovers and Biased Technical Change in Agriculture', *Economic Systems Research* 11, 31–48.

Verspagen, B. (1997), 'Estimating International Technology Spillovers Using Technology Flow Matrices', *Weltwirtschaftliches Archiv,* 133, 226–48.

Wolcott, S. (1994), 'The Perils of Lifetime Employment Systems: Productivity Advance in the Indian and Japanese Textile Industries, 1920–1938', *Journal of Economic History*, 54, 307–24.

Wood, A. (1994), *North–South Trade, Employment and Inequality: Changing Fortunes in a Skill-Driven World*, Clarendon Press, Oxford.

Wood, A. (1998), 'Globalisation and the Rise in Labour Market Inequalities', draft for A. Wood, (1998) 'Globalisation and the Rise in Labour Market Inequalities', *Economic Journal*, 108, 1463–82.

Worldbank (1997), *World Development Indicators CD rom*, Washington.

Yang, X. and B.J. Heijdra (1993), 'Monopolistic Competition and Optimum Product Diversity: Comment', *American Economic Review*, 83, 295–301.

Yorokoglu, M. (1998), 'The Information Technology Productivity Paradox', *Review of Economic Dynamics*, 1, 551–92.

Young, A. (1994), 'Lessons From the East Asian NICs: A Contrarian View', *European Economic Review, Papers and Proceedings,* 38, 964–73.

Young, A. (1995), 'The Tyranny of Numbers: Confronting the Statistical Realities of the East Asian Growth Experience', *Quarterly Journal of Economics*, 110, 641–80.

Young, A. (1998), 'Growth Without Scale Effects', *Journal of Political Economy*, 106, 41–63.

Index